RELATIVISTIC FIGURES OF EQUILIBRIUM

Ever since Newton introduced his theory of gravity, many famous physicists and mathematicians have worked on the problem of determining the properties of rotating bodies in equilibrium, such as planets and stars. In recent years, neutron stars and black holes have become increasingly important, and observations by astronomers and modelling by astrophysicists have reached the stage where rigorous mathematical analysis needs to be applied in order to understand their basic physics.

This book treats the classical problem of gravitational physics within Einstein's theory of general relativity. It begins by presenting basic principles and equations needed to describe rotating fluid bodies, as well as black holes in equilibrium. It then goes on to deal with a number of analytically tractable limiting cases, placing particular emphasis on the rigidly rotating disc of dust. The book concludes by considering the general case, using powerful numerical methods that are applied to various models, including the classical example of equilibrium figures of constant density.

Researchers in general relativity, mathematical physics and astrophysics will find this a valuable reference book on the topic. A related website containing codes for calculating various figures of equilibrium is available at www.cambridge.org/9780521863834.

REINHARD MEINEL is a Professor of Theoretical Physics at the Theoretisch-Physikalisches Institut, Friedrich-Schiller-Universität, Jena, Germany. His research is in the field of gravitational theory, focusing on astrophysical applications.

MARCUS ANSORG is a Researcher at the Max-Planck-Institut für Gravitationsphysik, Potsdam, Germany, where his research focuses on the application of spectral methods for producing highly accurate solutions to Einstein's field equations.

ANDREAS KLEINWÄCHTER is a Researcher at the Theoretisch-Physikalisches Institut, Friedrich-Schiller-Universität. His current research is on analytical and numerical methods for solving the axisymmetric and stationary equations of general relativity.

GERNOT NEUGEBAUER is a Professor Emeritus at the Theoretisch-Physikalisches Institut, Friedrich-Schiller-Universität. His research deals with Einstein's theory of gravitation, soliton theory and thermodynamics.

DAVID PETROFF is a Researcher at the Theoretisch-Physikalisches Institut, Friedrich-Schiller-Universität. His research is on stationary black holes and neutron stars, making use of analytical approximations and numerical methods.

RELATIVISTIC FIGURES OF EQUILIBRIUM

REINHARD MEINEL
Friedrich-Schiller-Universität, Jena

MARCUS ANSORG
Max-Planck-Institut für Gravitationsphysik, Potsdam

ANDREAS KLEINWÄCHTER
Friedrich-Schiller-Universität, Jena

GERNOT NEUGEBAUER
Friedrich-Schiller-Universität, Jena

DAVID PETROFF
Friedrich-Schiller-Universität, Jena

CAMBRIDGE UNIVERSITY PRESS
Cambridge, New York, Melbourne, Madrid, Cape Town, Singapore, São Paulo, Delhi

Cambridge University Press
The Edinburgh Building, Cambridge CB2 8RU, UK

Published in the United States of America by Cambridge University Press, New York

www.cambridge.org
Information on this title: www.cambridge.org/9780521863834

© R. Meinel, M. Ansorg, A. Kleinwächter, G. Neugebauer and D. Petroff 2008

This publication is in copyright. Subject to statutory exception
and to the provisions of relevant collective licensing agreements,
no reproduction of any part may take place without
the written permission of Cambridge University Press.

First published 2008

Printed in the United Kingdom at the University Press, Cambridge

A catalogue record for this publication is available from the British Library

Library of Congress Cataloguing in Publication data
Relativistic figures of equilibrium / Reinhard Meinel ... [et al.].
p. cm.
Includes bibliographical references and index.
ISBN 978-0-521-86383-4 (hardback : alk. paper)
1. Rotating masses of fluid. 2. Equilibrium. 3. Relativity (Physics) 4. Astrophysics.
I. Meinel, Reinhard. II. Title.
QB410.R45 2008
523–dc22   2008013352

ISBN 978-0-521-86383-4 hardback

Cambridge University Press has no responsibility for the persistence or
accuracy of URLs for external or third-party internet websites referred to
in this publication, and does not guarantee that any content on such
websites is, or will remain, accurate or appropriate.

Contents

Preface	*page*	vii
Notation		ix
1 **Rotating fluid bodies in equilibrium: fundamental notions and equations**		1
1.1 The concept of an isolated body		1
1.2 Fluid bodies in equilibrium		3
1.3 The metric of an axisymmetric perfect fluid body in stationary rotation		3
1.4 Einstein's field equations inside and outside the body		5
1.5 Equations of state		10
1.6 Physical properties		13
1.7 Limiting cases		16
1.8 Transition to black holes		26
2 **Analytical treatment of limiting cases**		34
2.1 Maclaurin spheroids		34
2.2 Schwarzschild spheres		38
2.3 The rigidly rotating disc of dust		40
2.4 The Kerr metric as the solution to a boundary value problem		108
3 **Numerical treatment of the general case**		114
3.1 A multi-domain spectral method		115
3.2 Coordinate mappings		128
3.3 Equilibrium configurations of homogeneous fluids		137
3.4 Configurations with other equations of state		153
3.5 Fluid rings with a central black hole		166
4 **Remarks on stability and astrophysical relevance**		177

Appendix 1 A detailed look at the mass-shedding limit   181
Appendix 2 Theta functions: definitions and relations   187
Appendix 3 Multipole moments of the rotating disc of dust   193
Appendix 4 The disc solution as a Bäcklund limit   203
References   208
Index   216

Preface

The theory of figures of equilibrium of rotating, self-gravitating fluids was developed in the context of questions concerning the shape of the Earth and celestial bodies. Many famous physicists and mathematicians such as Newton, Maclaurin, Jacobi, Liouville, Dirichlet, Dedekind, Riemann, Roche, Poincaré, H. Cartan, Lichtenstein and Chandrasekhar made important contributions. Within Newton's theory of gravitation, the shape of the body can be inferred from the requirement that the force arising from pressure, the gravitational force and the centrifugal force (in the corotating frame) be in equilibrium. Basic references are the books by Lichtenstein (1933) and Chandrasekhar (1969).

Our intention with the present book is to treat the *general relativistic* theory of equilibrium configurations of rotating fluids. This field of research is also motivated by astrophysics: neutron stars are so compact that Einstein's theory of gravitation must be used for calculating the shapes and other physical properties of these objects. However, as in the books mentioned above, which inspired this book to a large extent, we want to present the basic theoretical framework and will not go into astrophysical detail. We place emphasis on the rigorous treatment of simple models instead of trying to describe real objects with their many complex facets, which by necessity would lead to ephemeral and inaccurate models.

The basic equations and properties of equilibrium configurations of rotating fluids within general relativity are described in Chapter 1. We start with a discussion of the concept of an isolated body, which allows for the treatment of a single body without the need for dealing with the 'rest of the universe'. In fact, the assumption that the distant external world is *isotropic*, makes it possible to *justify* the condition of 'asymptotic flatness' in the body's far field region. Rotation 'with respect to infinity' then means nothing more than rotation with respect to the distant environment (the 'fixed stars') – very much in the spirit of Mach's principle. The main part of Chapter 1 provides a consistent mathematical formulation of the rotating fluid body problem within general relativity including its thermodynamic aspects. Conditions for parametric (quasi-stationary) transitions from rotating fluid bodies to black holes are also discussed.

Chapter 2 is devoted to the careful analytical treatment of limiting cases: (i) the Maclaurin spheroids, a well-known sequence of axisymmetric equilibrium configurations of homogeneous fluids in the Newtonian limit; (ii) the Schwarzschild spheres, representing non-rotating, relativistic configurations with constant mass-energy density; and (iii) the relativistic solution for a uniformly rotating disc of dust. The exact solution to the disc problem is rather involved and a detailed derivation of it will be provided here, which includes a discussion of aspects that have not been dealt with elsewhere. The solution is derived by applying the 'inverse method' – first used to solve the Korteweg–de Vries equation in the context of soliton theory – to Einstein's equations. The mathematical and physical properties of the disc solution including its black hole limit (extreme Kerr metric) are discussed in some detail. At the end of Chapter 2, we show that the inverse method also allows one to *derive* the general Kerr metric as the unique solution to the Einstein vacuum equations for well-defined boundary conditions on the horizon of the black hole.

In Chapter 3, we demonstrate how one can solve general fluid body problems by means of numerical methods. We apply them to give an overview of relativistic, rotating, equilibrium configurations of constant mass-energy density. Configurations with other selected equations of state as well as ring-like bodies with a central black hole are treated summarily. A *related website* provides the reader with, amongst other things, a computer code based on a highly accurate spectral method for calculating various equilibrium figures.

Finally, we discuss some aspects of stability of equilibrium configurations and their astrophysical relevance.

We hope that our book – with its presentation of analytical *and* numerical methods – will be of value to students and researchers in general relativity, mathematical physics and astrophysics.

Acknowledgments

Many thanks to Cambridge University Press for all its help during the preparation and production of this book. Support from the Dentsche Forschungsgemeinschaft through the Transregional Collaborative Research Centre 'Gravitational Wave Astronomy' is also gratefully acknowledged.

Notation

Units: $G = c = 1$ (G: Newton's gravitational constant, c: speed of light)

Complex conjugation: $\overline{a + ib} = a - ib$ (a, b real)

Greek indices ($\alpha, \beta, \ldots$): run from 1 to 3

Latin indices ($a, b, \ldots$): run from 1 to 4

Minkowski space: $ds^2 = \eta_{ab} dx^a dx^b = dx^2 + dy^2 + dz^2 - dt^2$
$$(x^1 = x, x^2 = y, x^3 = z, x^4 = t)$$

Metric of a rotating fluid body in equilibrium:

$$ds^2 = e^{-2U}\left[e^{2k}(d\varrho^2 + d\zeta^2) + W^2 d\varphi^2\right] - e^{2U}(dt + a\, d\varphi)^2$$
$$= e^{2\alpha}(d\varrho^2 + d\zeta^2) + W^2 e^{-2\nu}(d\varphi - \omega\, dt)^2 - e^{2\nu} dt^2$$

Killing vectors: $\boldsymbol{\xi} = \partial/\partial t$ and $\boldsymbol{\eta} = \partial/\partial \varphi$

Four-velocity of the fluid: $u^i = e^{-V}(\xi^i + \Omega\, \eta^i)$, $\quad \Omega =$ constant

Energy-momentum tensor: $T_{ik} = (\epsilon + p)\, u_i u_k + p\, g_{ik}$

Equation of state: $\epsilon = \epsilon(p)$

1
Rotating fluid bodies in equilibrium: fundamental notions and equations

1.1 The concept of an isolated body

An important and successful approach to solving problems throughout physics is to split the world into a system to be considered, its 'surroundings' and the 'rest of the universe', where the influence of the latter on the system being considered is neglected. The applicability of this concept to general relativity is not a trivial matter, since the spacetime structure at every point depends on the overall energy-momentum distribution.

Our aim is to find a description of a single fluid body (modelling a celestial body, e.g. a neutron star) under the influence of its own gravitational field. Fortunately, one often encounters such a body surrounded by a vacuum, where the closest other bodies are so far away that an intermediate region with a weak gravitational field exists. In such a situation (see Fig. 1.1) one can discuss the *far field* of the body. If the distant outside world (the 'rest of the universe') is *isotropic*, which it is according to astronomical observations and the standard cosmological models, then the line element corresponding to the far field of an arbitrary stationary body can be written as follows (see Stephani 2004):

$$ds^2 = g_{ab}dx^a dx^b = g_{\alpha\beta}dx^\alpha dx^\beta + 2g_{\alpha 4}dx^\alpha dt + g_{44}dt^2,$$

with

$$g_{\alpha\beta} = (1 + 2M/r)\eta_{\alpha\beta} + \mathcal{O}(r^{-2}),$$

$$g_{\alpha 4} = 2r^{-3}\epsilon_{\alpha\beta\gamma}x^\beta J^\gamma + \mathcal{O}(r^{-3}), \quad (1.1)$$

$$g_{44} = -(1 - 2M/r) + \mathcal{O}(r^{-2}),$$

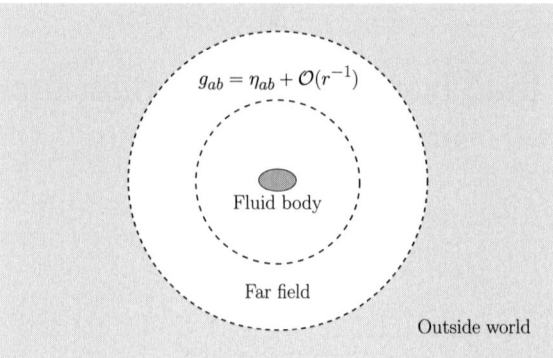

Fig. 1.1. The far field of an isolated body (adapted from Stephani 2004).

where $r^2 = \eta_{\alpha\beta}x^\alpha x^\beta = x^2+y^2+z^2$. For $r \to \infty$ the metric acquires the Minkowski form, i.e. the spacetime is 'asymptotically flat'. We stress that the condition of asymptotic flatness as discussed here is a consequence of the assumption of an isotropic outside world.[1]

M is the gravitational mass of the body and J^α its angular momentum. The $g_{\alpha 4}$-term represents the famous Lense–Thirring effect of a rotating source on the gravitational field, also called the 'gravitomagnetic' effect – in analogy to the magnetic field generated by a rotating electric charge distribution in Maxwell's electrodynamics.

In the next section, we shall provide arguments suggesting that the metric of a rotating fluid body in equilibrium is axially symmetric. Therefore, throughout this book, we shall deal with stationary and axisymmetric spacetimes. Under these conditions, the exterior (vacuum) Einstein equations can be reduced to the so-called Ernst equation, which can be attacked by analytic solution methods from soliton theory. However, the full rotating body problem requires the simultaneous solution of the inner equations, including the correct matching conditions. Note that the shape of the body's surface is not known in advance! The final result must be a globally regular and asymptotically flat solution to the Einstein equations, which can only be found by numerical methods in general (see Chapter 3). But, fortunately, there are a few interesting limiting cases that can be solved completely analytically (see Chapter 2).

---

[1] For an anisotropic outside world, it would be necessary to add a series with increasing powers of r to (1.1). The expressions (1.1), without these extra terms, would nevertheless be a good approximation to the body's far field as long as r is not too large ('local inertial system' on cosmic scales). However, for an isotropic outside world, the notion of a body's rotation with respect to the local inertial system coincides with the notion of rotation with respect to the external environment (the 'fixed stars'). Later, we shall simply speak of a rotation 'with respect to infinity'.

1.2 Fluid bodies in equilibrium

We want to consider configurations that are strictly stationary, thus implying thermodynamic equilibrium and the absence of gravitational radiation. This leads us, more or less stringently, to the conditions of

  (i) zero temperature,
 (ii) rigid rotation, and
(iii) axial symmetry.

Thermodynamic equilibrium would also permit a non-zero constant temperature.[2] However, as discussed for example in Landau and Lifshitz (1980), such configurations are unrealistic. Normal stars are hot, but not in global thermal equilibrium: their central temperature is much higher than their surface temperature and they emit a significant amount of electromagnetic radiation. Fortunately, neutron stars – the most interesting stars from the general relativistic point of view – can indeed be considered to be 'cold matter' objects, since their temperature is much lower than the Fermi temperature. Hence, our idealized assumption of zero temperature fits very well for neutron stars.

Provided that some (arbitrarily small) viscosity is present, any deviation from rigid rotation will vanish in an equilibrium state of a rotating star. For the calculation of the rigidly rotating equilibrium state itself, we may then adopt the model of a perfect fluid, since viscosity has no effect in the absence of any shear or expansion. It will, however, affect stability properties.

Moreover, within general relativity, any deviation of a uniformly rotating star from axial symmetry will result in gravitational radiation, which is also incompatible with a strict equilibrium state. For a more in-depth discussion of points (ii) and (iii), see Lindblom (1992).

Therefore, in the next sections, we shall treat stationary and axisymmetric, uniformly rotating, cold, perfect fluid bodies.

1.3 The metric of an axisymmetric perfect fluid body in stationary rotation

In accordance with our assumptions of axisymmetry and stationarity, we shall use coordinates t (time) and φ (azimuthal angle) adapted to the corresponding Killing vectors:

$$\xi = \frac{\partial}{\partial t}, \quad \eta = \frac{\partial}{\partial \varphi}, \qquad (1.2)$$

---

[2] Note that in general relativity, the equilibrium condition of constant temperature T is replaced by the Tolman condition $T(-g'_{tt})^{1/2} = $ constant (Tolman 1934), where the prime denotes a corotating frame of reference.

where $\boldsymbol{\xi}$ is normalized according to

$$\xi^i\xi_i \to -1 \quad \text{at spatial infinity} \tag{1.3}$$

and the orbits of the spacelike Killing vector $\boldsymbol{\eta}$ are closed, with periodicity 2π. The symmetry axis is characterized by

$$\boldsymbol{\eta} = 0 \quad \text{along the symmetry axis.} \tag{1.4}$$

It can be shown that the metric of an axisymmetric perfect fluid body in stationary rotation is orthogonally transitive, i.e. it admits 2-spaces orthogonal to the Killing vectors $\boldsymbol{\xi}$ and $\boldsymbol{\eta}$ (Kundt and Trümper 1966). This allows us to write the metric in the following form (Lewis 1932, Papapetrou 1966):

$$ds^2 = e^{-2U}\left[e^{2k}(d\varrho^2 + d\zeta^2) + W^2 d\varphi^2\right] - e^{2U}(dt + a\, d\varphi)^2, \tag{1.5}$$

or, equivalently,

$$ds^2 = e^{2\alpha}(d\varrho^2 + d\zeta^2) + W^2 e^{-2\nu}(d\varphi - \omega\, dt)^2 - e^{2\nu} dt^2, \tag{1.6}$$

where the functions U, a, k and W as well as ν, ω and α depend only on the coordinates ϱ and ζ. It can easily be verified that these functions are interrelated according to

$$\alpha = k - U, \quad W^{-1}e^{2\nu} \pm \omega = \left(W e^{-2U} \mp a\right)^{-1}. \tag{1.7}$$

We also note that U, a (or ν, ω) and W can be related to the scalar products of the Killing vectors, thus providing a coordinate independent characterization:

$$\xi^i\xi_i = -e^{2U} = -e^{2\nu} + \omega^2 W^2 e^{-2\nu}, \tag{1.8a}$$

$$\eta^i\eta_i = W^2 e^{-2U} - a^2 e^{2U} = W^2 e^{-2\nu}, \tag{1.8b}$$

$$\xi^i\eta_i = -a e^{2U} = -\omega W^2 e^{-2\nu}. \tag{1.8c}$$

We call U the 'generalized Newtonian potential' and a the 'gravitomagnetic potential'. Without loss of generality, the symmetry axis can be identified with the ζ-axis, i.e. it is characterized by $\varrho = 0$ and we have

$$0 \le \varrho < \infty, \quad -\infty < \zeta < \infty. \tag{1.9}$$

On the axis, the following conditions hold, see Stephani et al. (2003):

$$\varrho \to 0: \quad a \to 0,\ W \to 0,\ W/(\varrho\, e^k) \to 1. \tag{1.10}$$

At spatial infinity, i.e. for $\varrho^2+\zeta^2 \to \infty$, the line element approaches the Minkowski metric in cylindrical coordinates ϱ, ζ and φ:

$$ds^2 = d\varrho^2 + d\zeta^2 + \varrho^2 d\varphi^2 - dt^2, \tag{1.11}$$

which means that

$$U \to 0, \ a \to 0, \ k \to 0, \ W \to \varrho \quad \text{as} \quad \varrho^2 + \zeta^2 \to \infty \tag{1.12}$$

as well as

$$v \to 0, \ \omega \to 0, \ \alpha \to 0 \quad \text{as} \quad \varrho^2 + \zeta^2 \to \infty. \tag{1.13}$$

Sometimes we shall use a 'corotating coordinate system' characterized by

$$\varrho' = \varrho, \quad \zeta' = \zeta, \quad \varphi' = \varphi - \Omega t, \quad t' = t, \tag{1.14}$$

where Ω is the constant angular velocity of the fluid body with respect to infinity. It can easily be verified that the line element retains its form (1.5) or (1.6) with

$$e^{2U'} = e^{2U}[(1+\Omega a)^2 - \Omega^2 W^2 e^{-4U}], \tag{1.15a}$$

$$(1-\Omega a')e^{2U'} = (1+\Omega a)e^{2U}, \tag{1.15b}$$

$$k' - U' = k - U, \quad W' = W \tag{1.15c}$$

and

$$v' = v, \quad \omega' = \omega - \Omega, \quad \alpha' = \alpha. \tag{1.16}$$

Note that

$$\frac{\partial}{\partial t'} = \xi + \Omega \eta, \quad \frac{\partial}{\partial \varphi'} = \eta. \tag{1.17}$$

We shall call the primed quantities U', a', etc. 'corotating potentials'.

1.4 Einstein's field equations inside and outside the body

The stationary and rigid rotation of the fluid is characterized by the 4-velocity field

$$u^i = e^{-V}(\xi^i + \Omega \eta^i), \quad \Omega = \text{constant}, \tag{1.18}$$

where $\Omega = d\varphi/dt = u^\varphi/u^t$ is the constant angular velocity with respect to infinity. Using $u^i u_i = -1$, the factor $e^{-V} = u^t$ is given by

$$(\xi^i + \Omega \eta^i)(\xi_i + \Omega \eta_i) = -e^{2V}. \tag{1.19}$$

Note that V is equal to the corotating potential U',

$$V \equiv U', \tag{1.20}$$

as defined in (1.15a). The energy-momentum tensor of a perfect fluid is

$$T_{ik} = (\epsilon + p)\, u_i u_k + p\, g_{ik}, \tag{1.21}$$

where the mass-energy density ϵ and the pressure p, according to our assumptions as discussed in Section 1.2, are related by a 'cold' equation of state $\epsilon = \epsilon(p)$ following from

$$\epsilon = \epsilon(\mu_B), \quad p = p(\mu_B) \tag{1.22}$$

at zero temperature, with the baryonic mass-density μ_B. Examples will be given in Section 1.5.

The specific enthalpy[3]

$$h = \frac{\epsilon + p}{\mu_B} \tag{1.23}$$

can be calculated from $\epsilon(p)$ via the thermodynamic relation

$$dh = \frac{1}{\mu_B}\, dp \quad \text{(zero temperature)} \tag{1.24}$$

leading to

$$\frac{dh}{h} = \frac{dp}{\epsilon + p} \quad \Rightarrow \quad h(p) = h(0)\, \exp\left[\int_0^p \frac{dp'}{\epsilon(p') + p'}\right]. \tag{1.25}$$

Note that $h(0) = 1$ for ordinary baryonic matter.[4] From $T^{ik}{}_{;k} = 0$ (a semicolon denotes the covariant derivative), we obtain, as a first integral of the equations inside the body,

$$h(p)\, e^V = h(0)\, e^{V_0} = \text{constant}. \tag{1.26}$$

This means that surfaces of constant p coincide with surfaces of constant V. The boundary of the fluid body is defined by $p = 0$, hence

$$V = V_0 \quad \text{along the boundary of the fluid.} \tag{1.27}$$

---

[3] Note that $\epsilon = \mu_B + u_{\text{int}}$, where u_{int} denotes the internal energy-density. Hence $h = 1 + h_N$ with h_N being the specific enthalpy as it is usually defined in the non-relativistic (Newtonian) theory.
[4] An exception is strange quark matter as described by the MIT bag model, see Section 1.5.

The constant V_0 is related to the relative redshift z of zero angular momentum photons[5] emitted from the surface of the fluid and received at infinity via

$$z = e^{-V_0} - 1. \tag{1.28}$$

Equilibrium models, for a given equation of state, are fixed by two parameters, for example Ω and V_0.

The full set of equations that follows from Einstein's field equations $R_{ik} - \frac{1}{2}Rg_{ik} = 8\pi T_{ik}$ for the metric in the form (1.6), with (1.18) and (1.21), can be written in the following way, see e.g. Bardeen and Wagoner (1971):

$$\nabla \cdot (B\nabla \nu) - \frac{1}{2}\varrho^2 B^3 e^{-4\nu}(\nabla\omega)^2 = 4\pi e^{2\alpha}B\left[(\epsilon + p)\frac{1+v^2}{1-v^2} + 2p\right], \tag{1.29a}$$

$$\nabla \cdot (\varrho^2 B^3 e^{-4\nu}\nabla\omega) = -16\pi\varrho B^2 e^{2\alpha-2\nu}(\epsilon + p)\frac{v}{1-v^2}, \tag{1.29b}$$

$$\nabla \cdot (\varrho\nabla B) = 16\pi\varrho B e^{2\alpha}p \tag{1.29c}$$

with

$$B := W/\varrho \quad \text{and} \quad v := \varrho B e^{-2\nu}(\Omega - \omega), \tag{1.29d}$$

together with two equations, which provide the possibility of determining α via a line integral if the other three functions ν, ω and B are considered as given,

$$\varrho^{-1}(\alpha + \nu)_{,\varrho} + B^{-1}[B_{,\varrho}(\alpha + \nu)_{,\varrho} - B_{,\zeta}(\alpha + \nu)_{,\zeta}] - \frac{1}{2}\varrho^{-2}B^{-1}(\varrho^2 B_{,\varrho})_{,\varrho}$$
$$+ \frac{1}{2}B^{-1}B_{,\zeta\zeta} - (\nu_{,\varrho})^2 + (\nu_{,\zeta})^2 + \frac{1}{4}\varrho^2 B^2 e^{-4\nu}[(\omega_{,\varrho})^2 - (\omega_{,\zeta})^2] = 0, \tag{1.30a}$$

$$\varrho^{-1}(\alpha + \nu)_{,\zeta} + B^{-1}[B_{,\varrho}(\alpha + \nu)_{,\zeta} + B_{,\zeta}(\alpha + \nu)_{,\varrho}] - \frac{1}{2}\varrho^{-2}B^{-1}(\varrho^2 B_{,\zeta})_{,\varrho}$$
$$- \frac{1}{2}B^{-1}B_{,\varrho\zeta} - 2\nu_{,\varrho}\nu_{,\zeta} + \frac{1}{2}\varrho^2 B^2 e^{-4\nu}\omega_{,\varrho}\omega_{,\zeta} = 0, \tag{1.30b}$$

and (1.26), which allows us to express p and ϵ, via (1.25) and the equation of state, in terms of

$$e^V \equiv e^{U'} = e^{\nu}\sqrt{1-v^2}. \tag{1.31}$$

---

[5] Zero angular momentum means $\eta_i p^i = 0$ (p^i: 4-momentum of the photon), i.e. the (conserved) component of the orbital angular momentum with respect to the symmetry axis vanishes. In particular, this is satisfied for all photons emitted from the poles of a body of spheroidal topology, since η vanishes on the axis of symmetry. For other points on the surface, the condition $\eta_i p^i = 0$ places a restriction on the directions of emission.

In (1.29), the operator ∇ has the same meaning as in a Euclidean 3-space in which ϱ, ζ and φ are cylindrical coordinates. Note that v as defined in (1.29d) is the linear velocity of rotation with respect to 'locally non-rotating observers'.[6] Its invariant definition is given by

$$\frac{v}{\sqrt{1-v^2}} = \frac{\eta_i u^i}{\sqrt{\eta_k \eta^k}}. \tag{1.32}$$

In (1.30), we have made use of the comma notation for partial derivatives, e.g. $\partial v/\partial \varrho = v_{,\varrho}$. Note that instead of (1.30), the second order equation for α

$$\alpha_{,\varrho\varrho} + \alpha_{,\zeta\zeta} - \frac{1}{\varrho} v_{,\varrho} + \nabla v \left(\nabla v - B^{-1} \nabla B \right) - \frac{1}{4} \varrho^2 B^2 e^{-4v} (\nabla \omega)^2$$
$$= -4\pi e^{2\alpha} (\epsilon + p), \tag{1.33}$$

which follows from (1.29), (1.30) and (1.26), see Trümper (1967), can be used.

For the metric in the form (1.5), the equations take a simpler form if one uses the *corotating* potentials U', a', k' and W'. With $W' = W$, see (1.15c), they read

$$\nabla^2 U' - \frac{1}{\varrho} U'_{,\varrho} + \frac{\nabla U' \cdot \nabla W}{W} + \frac{e^{4U'}(\nabla a')^2}{2W^2} = 4\pi(\epsilon + 3p)e^{2k'-2U'}, \tag{1.34a}$$

$$(W^{-1} e^{4U'} a'_{,\varrho})_{,\varrho} + (W^{-1} e^{4U'} a'_{,\zeta})_{,\zeta} = 0, \tag{1.34b}$$

$$W_{,\varrho\varrho} + W_{,\zeta\zeta} = 16\pi p W e^{2k'-2U'} \tag{1.34c}$$

together with

$$W_{,\varrho} k'_{,\varrho} - W_{,\zeta} k'_{,\zeta} = \frac{1}{2}(W_{,\varrho\varrho} - W_{,\zeta\zeta}) + W[(U'_{,\varrho})^2 - (U'_{,\zeta})^2]$$
$$+ \frac{e^{4U'}}{4W}[(a'_{,\zeta})^2 - (a'_{,\varrho})^2], \tag{1.35a}$$

$$W_{,\zeta} k'_{,\varrho} + W_{,\varrho} k'_{,\zeta} = W_{,\varrho\zeta} + 2W U'_{,\varrho} U'_{,\zeta} - \frac{e^{4U'}}{2W} a'_{,\varrho} a'_{,\zeta} \tag{1.35b}$$

and (1.26).

---

[6] Locally non-rotating observers (also called 'zero angular momentum observers') have a 4-velocity field $u^i_{\text{zamo}} = e^{-\nu}(\xi^i + \omega \eta^i)$. They rotate with the angular velocity ω with respect to infinity, but their angular momentum $\eta_i u^i_{\text{zamo}}$ vanishes, see Bardeen et al. (1972). This provides a nice interpretation for the metric functions ω and ν.

1.4 Einstein's field equations inside and outside the body

The vacuum case: the Ernst equation

Outside the body, the source terms on the right hand sides of Equations (1.34) vanish. Equation (1.34c) becomes a two-dimensional Laplace equation:

$$W_{,\varrho\varrho} + W_{,\zeta\zeta} = 0. \tag{1.36}$$

By means of a conformal transformation in ϱ-ζ space, it is always possible to choose

$$W \equiv \varrho. \tag{1.37}$$

In these 'canonical Weyl coordinates' the remaining field equations, written down for the functions U, a and k, are[7]

$$\nabla^2 U = -\frac{e^{4U}}{2\varrho^2}(\nabla a)^2, \tag{1.38}$$

$$(\varrho^{-1}e^{4U}a_{,\varrho})_{,\varrho} + (\varrho^{-1}e^{4U}a_{,\zeta})_{,\zeta} = 0 \tag{1.39}$$

together with the two equations

$$k_{,\varrho} = \varrho[(U_{,\varrho})^2 - (U_{,\zeta})^2] + \frac{e^{4U}}{4\varrho}[(a_{,\zeta})^2 - (a_{,\varrho})^2], \tag{1.40a}$$

$$k_{,\zeta} = 2\varrho U_{,\varrho}U_{,\zeta} - \frac{e^{4U}}{2\varrho}a_{,\varrho}a_{,\zeta}, \tag{1.40b}$$

which allow us to calculate k via a path-independent[8] line integral.

Equation (1.39) implies that a function b can be introduced according to

$$a_{,\varrho} = \varrho e^{-4U}b_{,\zeta}, \quad a_{,\zeta} = -\varrho e^{-4U}b_{,\varrho}, \tag{1.41}$$

satisfying the equation

$$(\varrho e^{-4U}b_{,\varrho})_{,\varrho} + (\varrho e^{-4U}b_{,\zeta})_{,\zeta} = 0. \tag{1.42}$$

It can easily be verified that the two Equations (1.38) and (1.42) can be combined into the Ernst equation (Ernst 1968, Kramer and Neugebauer 1968).

$$\Re f \, \nabla^2 f = (\nabla f)^2 \tag{1.43}$$

---

[7] As a consequence of the form invariance of the line element (1.5) under a coordinate transformation (1.14), the vacuum equations for U, a, k and W are the same as those for U', a', k' and W' ($= W$), and can be read off from Equations (1.34) and (1.35) for $\epsilon = p = 0$.

[8] The integrability condition is satisfied by virtue of (1.38) and (1.39).

for the complex 'Ernst potential'

$$f := e^{2U} + ib. \quad (1.44)$$

The Ernst equation (1.43), together with (1.41), (1.40) and (1.37), is equivalent to the vacuum Einstein equations in the stationary and axisymmetric case.

As already mentioned, the vacuum equations for the corotating potentials U' and a' have the same form as those for U and a. Therefore, the Ernst potential can also be introduced in the corotating system and the Ernst equation retains its form as well. This remarkable fact will be used later.

The global problem

For genuine fluid body problems, we shall not make use of canonical Weyl coordinates and the Ernst formalism in the exterior region. It is of greater advantage to have a global coordinate system ϱ, ζ in which all metric functions and their first derivatives are continuous at the surface of the body. In particular, this requirement leads to a unique solution $W(\varrho, \zeta)$, which differs from $W \equiv \varrho$ in the vacuum region.[9] The global problem consists in finding a regular, asymptotically flat solution to Equations (1.29) and (1.30) with source terms inside the fluid and without source terms in the vacuum region. We stress that the shape of the surface, characterized by $p = 0$, is not known from the outset.

1.5 Equations of state

In this section, we shall provide some examples of equations of state $\epsilon = \epsilon(p)$, which will be used in this book. The relation to the baryonic mass-density μ_B, consistent with Equations (1.23) and (1.25), will also be given. Note that in our units (with $c = 1$), there is no difference between energy-density ϵ and (total) mass-density μ, i.e. $\epsilon = \mu = \mu_B + u_{\text{int}}$, where u_{int} is the internal energy-density.

Homogeneous fluids

This simple model is characterized by the equation of state (EOS)

$$\epsilon = \text{constant}. \quad (1.45)$$

Assuming $h(0) = 1$, we obtain from (1.23) and (1.25) that $\epsilon = \mu = \mu_B$, i.e. the internal energy density is zero.

---

[9] An important exception is given by the disc limit, where it turns out that $W \equiv \varrho$ holds globally, see Subsection 1.7.3. Another application of the Ernst formalism will be the derivation of the Kerr metric in Section 2.4.

1.5 Equations of state

Relativistic polytropes

This model is defined by

$$p = K\mu_B^\gamma, \quad \gamma = 1 + \frac{1}{n} \quad (n > 0), \tag{1.46}$$

see Tooper (1965). Here K is called the 'polytropic constant', γ the 'polytropic exponent', and n the 'polytropic index'. With $h(0) = 1$, we obtain from (1.23) and (1.25) the relation

$$\epsilon = \mu_B + np, \tag{1.47}$$

i.e. $p = (\gamma - 1)u_{\text{int}}$ and the EOS reads $\epsilon = (p/K)^{1/\gamma} + p/(\gamma - 1)$. Note that the homogeneous case $\epsilon = $ constant is contained as the limit $n \to 0$. It should, however, be noted that this EOS – if applied in the dynamic case – guarantees a speed of sound less than the speed of light only for $n \geq 1$.

Completely degenerate, ideal gas of neutrons

The general EOS for a completely degenerate, ideal Fermi gas (a genuine zero-temperature EOS!) was derived by Stoner (1932) in the framework of special-relativistic Fermi–Dirac statistics. The two limiting cases, the non-relativistic and ultra-relativistic limit, lead to polytropic relations (1.46) with exponents $\gamma = 5/3$ and $\gamma = 4/3$ respectively. This EOS, applied to an electron gas, plays a crucial role in the theory of white dwarfs, see Chandrasekhar (1939). Here we have in mind the application to a neutron gas, first considered by Landau (1932), and used in the famous work by Oppenheimer and Volkoff (1939) to calculate models of neutron stars. Pressure, energy-density and baryonic mass-density are related as follows:

$$p = \frac{m_n^4}{24\pi^2 \hbar^3} f(x), \tag{1.48a}$$

$$\epsilon = \mu_B + \frac{m_n^4}{24\pi^2 \hbar^3} g(x), \tag{1.48b}$$

$$\mu_B = \frac{m_n^4}{3\pi^2 \hbar^3} x^3, \tag{1.48c}$$

where

$$f(x) = x(2x^2 - 3)(x^2 + 1)^{1/2} + 3\operatorname{arcsinh} x,$$

$$g(x) = 8x^3 \left[(x^2 + 1)^{1/2} - 1 \right] - f(x),$$

see, for example, Kippenhahn and Weigert (1990). Here m_n is the mass of a neutron and $\hbar$ is the reduced Planck constant (remember that we use units with $c = 1$). It can easily be verified that $h(0) = 1$ and that (1.25) is satisfied.

Strange quark matter as described by the MIT bag model

This model, in its simplest version, leads to

$$\epsilon = A\mu_B^{4/3} + B, \quad p = \frac{1}{3}A\mu_B^{4/3} - B, \tag{1.49}$$

with well-defined constants A and B, see, for example, Gourgoulhon et al. (1999) and references therein (B is called the 'MIT bag constant'). The resulting EOS $\epsilon = \epsilon(p)$ is very simple:

$$\epsilon = 3p + 4B. \tag{1.50}$$

Again, the thermodynamic relation (1.25) is satisfied. Note that here

$$h(0) = \left.\frac{\epsilon}{\mu_B}\right|_{p=0} = 4B^{-1/4}(A/3)^{3/4} < 1, \tag{1.51}$$

corresponding to the assumption that strange quark matter (also called strange matter) represents the absolute ground state of matter at zero pressure and temperature.

The dust limit

The dust model is characterized by

$$p = 0, \tag{1.52}$$

i.e. the energy-momentum tensor reduces to

$$T_{ik} = \epsilon u_i u_k. \tag{1.53}$$

In this case, the equations $T^{ik}{}_{;k} = 0$ imply geodesic motion of the fluid elements,

$$u^i{}_{;k} u^k = 0, \tag{1.54}$$

and the local conservation law

$$(\epsilon u^i)_{;i} = 0, \tag{1.55}$$

which allows us to identify the mass-energy density with the baryonic mass-density,

$$\epsilon = \mu_B, \quad \text{i.e.} \quad h(p) = h(0) = 1. \tag{1.56}$$

1.6 Physical properties

1.6.1 Mass and angular momentum

The gravitational mass M and the total angular momentum J (strictly speaking, the component with respect to the axis of symmetry) in an asymptotically flat, stationary and axisymmetric spacetime are given by

$$M = 2\int_\Sigma (T_{ik} - \tfrac{1}{2}T^j{}_j g_{ik})n^i\xi^k \mathrm{d}\mathcal{V}, \quad J = -\int_\Sigma T_{ik}\, n^i\eta^k \mathrm{d}\mathcal{V}, \tag{1.57}$$

where Σ is a spacelike hypersurface ($t = $ constant) with the volume element $\mathrm{d}\mathcal{V} = \sqrt{{}^{(3)}g}\,\mathrm{d}^3x$ and the future pointing unit normal n^i, see for example Wald (1984). Note that $\eta^i n_i = 0$. The baryonic mass M_0, corresponding to the local conservation law $(\mu_B u^i)_{;i} = 0$, is given by the expression

$$M_0 = -\int_\Sigma \mu_B\, u_i\, n^i \mathrm{d}\mathcal{V}. \tag{1.58}$$

Nearby equilibrium configurations with the same equation of state are related by

$$\delta M = \Omega\, \delta J + \mu_c\, \delta M_0, \quad \mu_c = h(0)\mathrm{e}^{V_0}. \tag{1.59}$$

This follows from a variational principle (Hartle and Sharp 1967, see also Bardeen 1970 and Neugebauer 1988). The factor $\mu_c = h(0)\mathrm{e}^{V_0}$ thus plays the role of the equilibrium value of the body's chemical potential (in appropriate units):

$$\mu_c = \left.\frac{\partial M}{\partial M_0}\right|_{J=\text{constant}}. \tag{1.60}$$

Note that

$$h(p)\,\mathrm{e}^V = \mu_c = \text{constant}, \tag{1.61}$$

see (1.26), is indeed the Tolman equilibrium relation for the chemical potential h.[10]

The gravitational mass M and the total angular momentum J can also be read off from the asymptotic behaviour of $\xi^i\xi_i$ and $\xi^i\eta_i/\eta^k\eta_k$ as $r = \sqrt{\varrho^2 + \zeta^2} \to \infty$:

$$\xi^i\xi_i = -1 + \frac{2M}{r} + \mathcal{O}(r^{-2}), \tag{1.62}$$

$$\frac{\xi^i\eta_i}{\eta^k\eta_k} = -\frac{2J}{r^3} + \mathcal{O}(r^{-4}), \tag{1.63}$$

---

[10] In the zero temperature case, there is no difference between enthalpy and free enthalpy (Gibbs free energy).

which, for the metric as given in (1.5) or (1.6), means[11]

$$U = -\frac{M}{r} + \mathcal{O}(r^{-2}), \quad a = \frac{2J\varrho^2}{r^3} + \mathcal{O}(r^{-2}) \tag{1.64}$$

or, equivalently,

$$\nu = -\frac{M}{r} + \mathcal{O}(r^{-2}), \quad \omega = \frac{2J}{r^3} + \mathcal{O}(r^{-4}). \tag{1.65}$$

In asymptotically Cartesian coordinates $x^1 = \varrho \cos \varphi$, $x^2 = \varrho \sin \varphi$, $x^3 = \zeta$, this corresponds exactly to (1.1) with $J^1 = J^2 = 0$, $J^3 = J$, where one has to take into account that $\alpha = -\nu + \mathcal{O}(r^{-2})$, which follows from (1.30).

1.6.2 Ergospheres

Regions in which the Killing vector ξ^i, which is normally timelike ($\xi^i \xi_i < 0$), becomes spacelike ($\xi^i \xi_i > 0$), are called *ergospheres*. The boundary of an ergosphere, also called an *ergosurface*, is characterized by $\xi^i \xi_i = 0$. Ergospheres appear when a rotating source becomes sufficiently relativistic, i.e. far away from the Newtonian limit. Despite the fact that the Killing vector ξ^i, corresponding to stationarity, is spacelike within the ergosphere, the spacetime can still be considered to be locally stationary, provided there exists a timelike linear combination of ξ^i and η^i. We can assume that the spacetime of rotating fluid bodies in equilibrium is locally stationary everywhere. It should be noted, however, that this condition is violated inside, and on the event horizon of, black holes.

Next we discuss some mathematical and physical aspects of the presence of ergospheres.

Mathematical aspects

For the metric in the form

$$ds^2 = e^{2\alpha}(d\varrho^2 + d\zeta^2) + W^2 e^{-2\nu}(d\varphi - \omega \, dt)^2 - e^{2\nu} dt^2, \tag{1.66}$$

nothing special happens with α, W, ν and ω at an ergosurface[12] or inside the ergosphere. The local stationarity of the spacetime is guaranteed precisely when $e^{2\nu} > 0$, i.e. the function ν remains real inside the ergosphere. However, if we write the metric in the equivalent form

$$ds^2 = e^{-2U}\left[e^{2k}(d\varrho^2 + d\zeta^2) + W^2 d\varphi^2\right] - e^{2U}(dt + a \, d\varphi)^2, \tag{1.67}$$

---

[11] Note that $W = \varrho + \mathcal{O}(r^{-1})$ as $r \to \infty$.
[12] Special attention has to be paid to points at which the ergosurface reaches the symmetry axis. This happens at the poles of black hole horizons. For rotating fluid bodies, the ergosphere has a toroidal shape in general, and does not touch the rotation axis except for the (singular) limiting case of infinite central pressure.

we have to note that the function $e^{2U} = -\xi^i\xi_i$ changes its sign. Inside the ergosphere, $e^{2U} < 0$ holds, i.e. the function U is no longer real. The ergosurface is characterized by $e^{2U} = 0$. The behaviour of e^{2k} and a compensates for the 'dangerous' effects of $e^{2U} \leq 0$ such that all metric coefficients behave perfectly well at the ergosurface and inside the ergosphere.[13] In the vacuum case, one can introduce canonical Weyl coordinates (characterized by $W \equiv \varrho$) and the Ernst potential $f \equiv e^{2U} + ib$, see Section 1.4. It is important to note that f behaves perfectly well at the ergosurface, too. This means that the function b, in contrast to a, behaves well. Vice versa, analytic solutions of the Ernst equation (i.e. solutions for which e^{2U} and b can both be Taylor-expanded in the two real variables ϱ and ζ) with zero-level sets of e^{2U} lead to smooth ergosurfaces in spacetime, see Chruściel et al. (2006). Note that we continue to use the notation e^{2U} for $\Re f$ independent of its sign.

Physical aspects

The ergosurface is sometimes called the 'limiting surface of stationarity' or simply the 'stationary limit' since no timelike world lines with a tangent vector (4-velocity) proportional to ξ^i, which would represent observers that are stationary with respect to infinity, can exist any longer. Inside the ergosphere, the only term in the above line element ds^2, which may become negative, is the term $2g_{\varphi t}d\varphi dt$. Therefore, a timelike world line ($ds^2 < 0$) requires $d\varphi/dt \neq 0$, which means that observers *must rotate* about the axis of symmetry. The direction of this rotation is dictated by the sign of $g_{\varphi t} = -\omega W^2 e^{-2\nu}$:

$$\omega \frac{d\varphi}{dt} > 0 \quad \text{for timelike world lines within the ergosphere.} \tag{1.68}$$

For a uniformly rotating source, the sign of the function ω always coincides with the sign of the angular velocity Ω of the source, i.e. any observer must rotate in the same direction as the source inside the ergosphere.

It is interesting to note that the Killing vector $\partial/\partial t' = \boldsymbol{\xi} + \Omega \boldsymbol{\eta}$ of the corotating system, see Section 1.3, becomes spacelike *far away* from the rotation axis:

$$e^{2U'} = -(\xi^i + \Omega \eta^i)(\xi_i + \Omega \eta_i) \to -\Omega^2 \varrho^2 \quad \text{as} \quad \varrho \to \infty. \tag{1.69}$$

This corresponds to the well-known fact that no observers that are too distant from the axis can be stationary with respect to the corotating system as this would require superluminal motion.

---

[13] At the ergosurface, e^{2k} vanishes and a diverges. Inside the ergosphere, $e^{2k} < 0$.

1.7 Limiting cases

The few analytical solutions that can be found for figures of equilibrium rely on the fact that the problem simplifies significantly in certain limiting cases: (i) the Newtonian limit, where one has only a single gravitational potential satisfying the simple Poisson equation; (ii) the non-rotating limit, where the spherical symmetry implies a simple system of ordinary differential equations; and (iii) the disc limit, where a boundary value problem to the vacuum equations can be formulated. In this section we shall derive the relevant equations. The first two limits are well known, which enables us to be brief. The disc limit will be treated in much greater detail, since it is less well known and plays an important role in Chapter 2. In addition, a limiting case of a different nature, namely the mass-shedding limit, is also discussed. This limit poses particular challenges to the numerical methods to be presented in Chapter 3.

1.7.1 The Newtonian limit

The Newtonian limit, in our context, is approached when the following conditions are satisfied:

(i) The metric deviates only slightly from the Minkowski metric.
(ii) The linear velocity of rotation v, as defined in (1.32), is small as compared with the velocity of light: $v \ll c$, i.e. $v \ll 1$ in our units.
(iii) The pressure is small as compared with the mass-energy density: $p \ll \epsilon = \mu c^2$, i.e. $p \ll \mu$ in our units.[14]

It turns out that these conditions are all satisfied for rotating fluid bodies in equilibrium whenever the absolute value of the parameter V_0 becomes sufficiently small:

$$|V_0| \ll 1. \tag{1.70}$$

The metric function U becomes the Newtonian potential[15] satisfying the Poisson equation

$$\nabla^2 U = 4\pi\mu, \tag{1.71}$$

which reduces to the Laplace equation in the vacuum region:

$$\nabla^2 U = 0 \quad \text{outside the body.} \tag{1.72}$$

---

[14] This condition implies via (1.23) and (1.25) that $\mu \approx h(0)\mu_B$, i.e. for all equations of state with $h(0) = 1$, the mass-density μ can be identified with the baryonic mass-density μ_B in the Newtonian limit.
[15] Note that $\xi^i \xi_i = g_{tt} = -e^{2U} \approx -(1 + 2U) \approx -(1 + 2v)$ in the Newtonian limit.

The leading order terms of (1.25) and (1.26) give the Newtonian relation

$$V = V_0 - \int_0^p \frac{dp'}{\mu(p')}, \tag{1.73}$$

which is nothing other than the integrated form of the Euler equation

$$\nabla V = -\frac{\nabla p}{\mu} \tag{1.74}$$

with

$$V = U - \frac{1}{2}\Omega^2 \varrho^2. \tag{1.75}$$

The latter relation follows from (1.15a) and (1.20) to leading order. In a sense, V can be considered to be a 'corotating potential' in Newtonian theory as well (it includes the 'centrifugal potential' $-\Omega^2 \varrho^2/2$). Note, however, that V satisfies the equation

$$\nabla^2 V = 4\pi\mu - 2\Omega^2, \tag{1.76}$$

which does not reduce to the Laplace equation (1.72) outside the body. This is in remarkable contrast to the fact, discussed in Section 1.4, that the Ernst equation retains its form in the corotating system (a nice justification for calling Einstein's theory 'general relativity').

From (1.73) and $p = 0$, we obtain the Newtonian surface condition

$$V = V_0 \quad \text{along the surface}, \tag{1.77}$$

just as in general relativity.

1.7.2 The non-rotating limit

If one considers static, non-rotating fluid configurations, the field equations take a particularly simple form. Besides the mathematical simplification, this assumption is often justified on physical grounds, since most celestial bodies possess rather small rotation rates and hence a static model is a good approximation.

Since the spacetime continuum of a static perfect fluid body is spherically symmetric (see Masood-ul-Alam 2007), a corresponding form of the line element is appropriate. The field equations presented in Section 1.4 become ordinary differential equations with respect to the radial coordinate r that can be introduced alongside ϑ by $\varrho = r\sin\vartheta$, $\zeta = r\cos\vartheta$. The line element reads

$$ds^2 = e^{2\alpha}(dr^2 + r^2 d\vartheta^2 + r^2 \sin^2\vartheta\, d\varphi^2) - e^{2\nu} dt^2, \tag{1.78}$$

i.e. $\omega = 0$ and $W = \varrho e^{\alpha+\nu}$, cf. (1.6). These two conditions correspond to staticity and spherical symmetry respectively. However, in order to achieve a particularly concise form, one usually considers the field equations in standard Schwarzschild coordinates $(\tilde{r}, \vartheta, \varphi, t)$:

$$ds^2 = e^{2\tilde{\alpha}} d\tilde{r}^2 + \tilde{r}^2(d\vartheta^2 + \sin^2\vartheta \, d\varphi^2) - e^{2\nu} dt^2. \tag{1.79}$$

These coordinates are obtained through the simple, purely radial transformation

$$\tilde{r} = re^{\alpha}. \tag{1.80}$$

Since the matter is at rest in this coordinate system, we may write

$$u^i = (0, 0, 0, e^{-V}) \tag{1.81}$$

with

$$V = \nu. \tag{1.82}$$

Taking the integrated relativistic Euler equation (1.26) into account, one may derive the *Tolman–Oppenheimer–Volkoff* equation (Tolman 1939, Oppenheimer and Volkoff 1939):

$$\frac{dp}{d\tilde{r}} = \frac{(\epsilon + p)(m + 4\pi p \tilde{r}^3)}{\tilde{r}(2m - \tilde{r})}, \tag{1.83}$$

through which a 'mass function' $m(\tilde{r})$,

$$\frac{dm}{d\tilde{r}} = 4\pi \tilde{r}^2 \epsilon, \qquad m(0) = 0, \tag{1.84}$$

is determined. Equation (1.84) provides a relation between this function and the thermodynamic quantities ϵ, μ_B, p and h [see also Equations (1.22)–(1.25)], and by virtue of (1.26) and (1.82), the metric potential ν is also given. Moreover, the function $\tilde{\alpha}$ is obtained through

$$e^{-2\tilde{\alpha}} = 1 - \frac{2m}{\tilde{r}}. \tag{1.85}$$

If for a static perfect fluid model, the equation of state and a physical parameter (e.g. the central pressure p_c) are specified, then the complete interior solution can be determined through the above equations. The spatial location $\tilde{r} = \tilde{r}_0$ of the body's surface is then given by the condition of vanishing pressure. The metric in the exterior of the body is, of course, given by the well-known Schwarzschild vacuum solution, with the gravitational mass $M = m(\tilde{r}_0)$. Note that the constant of integration in (1.26) is then also fixed upon demanding continuity of the metric.

An important consequence, which can be derived under the reasonable assumption that the energy-density does not increase outwards, is the so-called *Buchdahl limit* (Buchdahl 1959): A spherically symmetric star can only exist in a state of equilibrium (can only compensate its own gravitational attraction with a finite pressure) if the ratio of its mass M to its radius $\tilde{r}_0$ satisfies the inequality

$$\frac{M}{\tilde{r}_0} < \frac{4}{9}. \tag{1.86}$$

Here the stellar radius $\tilde{r}_0$ assumes a coordinate invariant meaning by virtue of the relation

$$S = 4\pi \tilde{r}_0^2, \tag{1.87}$$

where S is the surface area of the star. The inequality (1.86) shows that a spherical star in equilibrium always has a (coordinate) radius greater than 9/8 times the Schwarzschild radius $2M$ of a black hole of the same mass. Beyond this limit, the star must inevitably collapse.

1.7.3 The disc limit

As a rule, a perfect fluid ball set in rotational motion takes on an oblate shape and we may expect that there are extremely flattened fluid configurations represented by an infinitely thin circular disc rotating about an axis of symmetry (in our context denoted by the ζ-axis). Here we shall construct a corresponding mathematical model. Later on, we shall show that the field equations are rigorously solvable in this limiting case (see Section 2.3). Exact solutions like this help to achieve deeper insight into the geometrical structure of the gravitational field of rotating bodies, facilitate a reliable discussion of physical effects and provide us with the interrelationship between characteristic parameters such as angular velocity, mass and angular momentum. Moreover, the study of the disc limit has astrophysical relevance: Discs play an important role as galaxy models or intermediate states in collapse processes. It should be mentioned that an approximate solution to the disc problem was found (Bardeen and Wagoner 1971) by solving a post-Newtonian expansion to high order numerically. The exact solution (Neugebauer and Meinel 1995) confirmed many of the predictions made in this notable paper.

The idea of the subsequent analysis is to describe the disc limit of perfect fluids by a boundary value problem of Einstein's *vacuum* equations with boundary data derived from the field equations *inside* the body, which degenerates to a circular disc with the coordinate radius ϱ_0 covering the domain $0 \leq \varrho \leq \varrho_0$ of a three-dimensional slice through spacetime – the 3-surface $\zeta = 0$, see Fig. 2.3.[16] This domain can be considered to be the world tube of the surface elements of the

---

[16] Unessential coordinates t and φ are omitted.

two-dimensional surface Σ_2,

$$\Sigma_2: \quad \zeta = 0 \quad (0 \leq \varrho \leq \varrho_0), \quad t = \text{constant},$$

in which energy and momentum of the fluid source are distributed as finite quantities per unit area in complete analogy to surface charge and surface current in electrodynamics. Hence, we can follow the treatment of surface layers in electrodynamics and derive junction conditions across the two-dimensional surface by integrating the field equations (1.34) over a 'pill box' that is centred on the surface and applying Gauss' theorem. These junction conditions combined with the obvious reflectional symmetry of the metric coefficients in (1.5),

$$\begin{aligned} U(\varrho, \zeta) &= U(\varrho, -\zeta), & a(\varrho, \zeta) &= a(\varrho, -\zeta), \\ k(\varrho, \zeta) &= k(\varrho, -\zeta), & W(\varrho, \zeta) &= W(\varrho, -\zeta), \end{aligned} \quad (1.88)$$

or, alternatively, in (1.6), will form the desired boundary conditions on the disc. In the limit $\zeta \to 0$, Equations (1.88) ensure the continuity of the metric across $\zeta = 0$ (including the surface layer),

$$\{U, a, k, W\}|_{\zeta=0^+} = \{U, a, k, W\}|_{\zeta=0^-} = \{U(\varrho, 0), a(\varrho, 0), k(\varrho, 0), W(\varrho, 0)\}, \quad (1.89)$$

and imply a jump in its normal derivative,

$$\{U_{,\zeta}, a_{,\zeta}, k_{,\zeta}, W_{,\zeta}\}|_{\zeta=0^+} = -\{U_{,\zeta}, a_{,\zeta}, k_{,\zeta}, W_{,\zeta}\}|_{\zeta=0^-}, \quad (1.90)$$

where $\zeta = 0^\pm$ means '$\zeta \to 0$ from above ($\zeta > 0$)' and '$\zeta \to 0$ from below ($\zeta < 0$)'. Note that Equations (1.88), (1.89) and (1.90) hold for the corotating potentials $\{U', a', k', W'\}$ too.

Before inspecting the field equations (1.34), we have to be aware of the behaviour of p and ϵ at the disc-like surface layer. We assume

$$p = \begin{cases} \text{finite on } \Sigma_2 \\ 0 \text{ outside } \Sigma_2; \end{cases} \quad \epsilon = \sigma_0(\varrho)\,\delta(\zeta), \quad (1.91)$$

where $\delta(\zeta)$ is the Dirac delta distribution. To motivate the first assumption, let us consider a (geometrical) transition from an oblate spheroidal fluid body (e.g. a Maclaurin ellipsoid) to a disc-like surface layer ('Maclaurin disc'). During all steps of the flattening process, the central (maximum) pressure should remain finite and the pressure retain its value, zero, on the body's surface. Consequently, any *volume* integral over p has to vanish in the disc limit ('set of measure zero'). Obviously, the volume energy-density ϵ becomes δ-infinite when related to surface elements of Σ_2.

1.7 Limiting cases

Taking into account Equation (1.91), the 'pill-box integration' of the field equation (1.34c) results in the junction condition

$$W_{,\zeta}|_{\zeta=0^+} - W_{,\zeta}|_{\zeta=0^-} = 0. \tag{1.92}$$

Using Equation (1.90) we arrive at the boundary condition

$$\Sigma_2: \quad W_{,\zeta}|_{\zeta=0^\pm} = 0 \tag{1.93}$$

for the vacuum equation (1.36)

$$W_{,\varrho\varrho} + W_{,\zeta\zeta} = 0.$$

Consider its reformulation

$$(W - \varrho)_{,\varrho\varrho} + (W - \varrho)_{,\zeta\zeta} = 0 \tag{1.94}$$

under the boundary conditions (1.93), (1.10) and (1.12), i.e.

$$\begin{aligned}(W - \varrho)_{,\zeta}|_{\zeta=0^\pm} &= 0 \quad \text{at} \quad \Sigma_2, \\ (W - \varrho) &\to 0 \quad \text{as} \quad \varrho \to 0 \quad \text{and also as} \quad \varrho^2 + \zeta^2 \to \infty.\end{aligned} \tag{1.95}$$

Obviously, the only regular solution $W - \varrho = W(\varrho, \zeta) - \varrho$ to this 'mixed' boundary value problem of the two-dimensional Laplace equation (1.94) is

$$W - \varrho = 0 \tag{1.96}$$

for all values of $\varrho \geq 0$ and ζ, i.e. we are automatically led to 'canonical Weyl coordinates', see Equation (1.37). From now on we set $W = \varrho$.

Since Equation (1.34b) has the form of a vanishing divergence, $\nabla \cdot (\varrho^{-2} e^{4U'} \nabla a') = 0$, we obtain, after a 'pill-box integration' and using (1.89) and (1.90),

$$\zeta = 0: \quad a'_{,\zeta}|_{\zeta=0} = 0 \tag{1.97}$$

in all points of the disc 'plane' $\zeta = 0$ including the disc-like surface layer. In analogy to (1.41), we may introduce a function b' via

$$a'_{,\varrho} = \varrho e^{-4U'} b'_{,\zeta}, \quad a'_{,\zeta} = -\varrho e^{-4U'} b'_{,\varrho} \tag{1.98}$$

satisfying

$$(\varrho e^{-4U'} b'_{,\varrho})_{,\varrho} + (\varrho e^{-4U'} b'_{,\zeta})_{,\zeta} = 0 \tag{1.99}$$

inside and outside the disc. Using (1.98) we get from (1.97)

$$\zeta = 0: \quad b'_{,\varrho}\bigg|_{\zeta=0} = 0 \quad \Rightarrow \quad b'(\varrho, 0) = \text{constant}, \tag{1.100}$$

where the arbitrary constant of integration can be put equal to zero,

$$b'(\varrho, 0) = 0. \tag{1.101}$$

This condition holds at all points of the disc 'plane' $\zeta = 0$, $t = $ constant inside and outside the disc layer.

Finally, we obtain from (1.34a)

$$\Sigma_2: \quad \sigma \equiv \sigma_0 e^{2k-2U} = \frac{1}{4\pi}\left(U'_{,\zeta}\big|_{\zeta=0^+} - U'_{,\zeta}\big|_{\zeta=0^-}\right) = \frac{1}{2\pi} U'_{,\zeta}\big|_{\zeta=0^+} \tag{1.102}$$

as a result of the 'pill-box integration', exploiting properties (1.91) of the matter distribution and the symmetry relations (1.89) and (1.90). Note that we have used $k' - U' = k - U$, see (1.15c). The surface energy-density σ_0, introduced in (1.91), as well as σ as defined in (1.102), depend on the choice of coordinates. An invariant ('proper') surface energy-density σ_p can be defined by

$$\Sigma_2: \quad \sigma_p = \frac{1}{2\pi} U'_{,i} N^i\bigg|_{\zeta=0^+} = \frac{1}{2\pi} U'_{,\zeta}\big|_{\zeta=0^+} e^{U-k}, \tag{1.103}$$

where $N^i = (\zeta_{,k}\zeta^{,k})^{-1/2}\zeta^{,i}$ is the unit normal vector of the timelike hypersurface $\zeta = 0$. Note that $N^i = e^{U-k}\delta^i_\zeta$ in the coordinates used here. Thus we can rewrite the volume energy-density in (1.91) to read

$$\epsilon = \sigma_p e^{U-k} \delta(\zeta). \tag{1.104}$$

To complete the boundary conditions, we recall that $U' = V$ (1.20) has to be constant on the surface of every fluid body, $V = V_0$, see (1.27). Therefore, we have to prescribe

$$\Sigma_2: \quad U' = V_0, \tag{1.105}$$

i.e. U' must be a constant in the disc layer. Because of the specific form of the boundary values (1.101) and (1.105), we choose the Ernst form of the vacuum equations (1.43) and describe the disc limit of perfect fluids by the boundary value problem

$$\Re f \nabla^2 f = (\nabla f)^2 \tag{1.106}$$

with

$$\zeta = 0, \quad \varrho \leq \varrho_0: \quad f'(\varrho, 0) = e^{2V_0}. \tag{1.107}$$

1.7 Limiting cases

Clearly, a regular solution of the boundary value problem has to satisfy the conditions (1.12) at spatial infinity. In terms of the Ernst potential f, they take the simple form

$$f \to 1 \quad \text{as} \quad \varrho^2 + \zeta^2 \to \infty. \tag{1.108}$$

For an illustration of the boundary values see Fig. 2.3. The mixture of primed and unprimed boundary values will lead to our using the 'corotating' Ernst equation

$$\Re f' \, \nabla^2 f' = (\nabla f')^2 \tag{1.109}$$

in the analysis of the boundary value problem as well.

To complete the formulae for the metric coefficients in (1.5), we go back to Section 1.4. Combining the vacuum relations (1.39) and (1.40) we have

$$\begin{aligned}
a_{,\varrho} &= \varrho e^{-4U} b_{,\zeta}, \quad a_{,\zeta} = -\varrho e^{-4U} b_{,\varrho}, \\
k_{,\varrho} &= \varrho \left\{ (U_{,\varrho})^2 - (U_{,\zeta})^2 + \frac{1}{4} e^{-4U} [(b_{,\varrho})^2 - (b_{,\zeta})^2] \right\}, \\
k_{,\zeta} &= 2\varrho \left\{ U_{,\varrho} U_{,\zeta} + \frac{1}{4} e^{-4U} b_{,\varrho} b_{,\zeta} \right\}.
\end{aligned} \tag{1.110}$$

Thus we can compute $a(\varrho, \zeta)$ as well as $k(\varrho, \zeta)$ from the Ernst function $f(\varrho, \zeta)$ via a path-independent line integration through the vacuum region, starting, say, at the axis of symmetry with the values $a = 0$ and $k = 0$ and ending at any point $\varrho \geq 0$, ζ including the points of the disc.

One expects that the reflectional symmetry will simplify the discussion of the boundary value problem. In terms of the Ernst functions $f = e^{2U} + ib$ and $f' = e^{2U'} + ib'$, we have

$$f(\varrho, -\zeta) = \overline{f(\varrho, \zeta)}, \quad f'(\varrho, -\zeta) = \overline{f'(\varrho, \zeta)}, \tag{1.111}$$

whence

$$\zeta = 0: \quad \bar{f}_{,\zeta}\big|_{\zeta=0^+} = -f_{,\zeta}\big|_{\zeta=0^-}, \quad \bar{f}'_{,\zeta}\big|_{\zeta=0^+} = -f'_{,\zeta}\big|_{\zeta=0^-}. \tag{1.112}$$

These relations can easily be derived from the corresponding relations (1.88)–(1.90) and the 'definition' of b and b' in (1.41) and (1.98) respectively. Note that the derivation implies a suitable choice of the integration constants, see (1.100), (1.101).

The previous analysis of the boundary value problem has shown that the metric of the disc solution must be continuous everywhere (even across the disc). However, the jumps in the first derivatives of the metric coefficients U, a, k across the disc, see (1.90), require a careful 'interpretation' of the state variables of the disc. We have already discussed the interrelation between the jump in the derivative $U'_{,\zeta}$ of the corotating 'generalized Newtonian potential' $U' = V$ [which is a function of U and a, see (1.15a)], the δ-like distribution of the energy-density ϵ and the 'smooth' behaviour of the pressure p, see Equations (1.91), (1.102)–(1.104). As a consequence, we may calculate the gravitational mass M and the total angular momentum J of the disc via (1.57) from the energy-momentum tensor

$$T_{ik} = \sigma_p\, e^{U-k} \delta(\zeta) u_i u_k, \tag{1.113}$$

where we have omitted the pressure term of (1.21), since a finite pressure cannot contribute to volume integrals over a surface layer. One has to keep this in mind when denoting (1.113) as the 'energy-momentum tensor of the disc'.

The 4-*velocity* u^i of the disc matter is well defined by (1.18), since its components $u^i \xi_i$ and $u^i \eta_i$ can be expressed in terms of the *continuous* metric coefficients U and a, see Equations (1.8), (1.20) and (1.15a). The definition of the 4-*acceleration* is based on the existence of the Christoffel symbols (first derivatives of the metric), which are not defined across the disc. A more general approach could start with the expression

$$v^i := e^{-V}(\xi^i + \Omega \eta^i), \quad v^i v_i = -1, \quad \Omega = \text{constant}, \tag{1.114}$$

which is well defined outside the disc. Interpreting (1.114) as the 4-velocity field of a cloud of particles ('observers'), we get for the 4-acceleration

$$\dot{v}_i \equiv v_{i;k} v^k = V_{,i}. \tag{1.115}$$

Obviously, v^i and u^i coincide along the surface layer. The same holds for the ϱ-components of the accelerations $\dot{v}_\varrho$ and $\dot{u}_\varrho$, which, according to (1.105), vanish along the surface layer,

$$\Sigma_2: \quad \dot{u}_\varrho = \dot{v}_\varrho = V_{,\varrho} = 0. \tag{1.116}$$

Interpreting the ζ-component of the 4-acceleration $\dot{u}_\zeta$ as the 'mean value' of the components $\dot{v}_\zeta$ 'from above' and 'from below',

$$\Sigma_2: \quad \dot{u}_\zeta := \frac{1}{2}\left(\dot{v}_\zeta\big|_{\zeta=0^+} + \dot{v}_\zeta\big|_{\zeta=0^-}\right) \tag{1.117}$$

and making use of $U'_{,\zeta}\big|_{\zeta=0^+} = -U'_{,\zeta}\big|_{\zeta=0^-}$, we may assert

$$\Sigma_2: \quad \dot{u}^i = 0, \tag{1.118}$$

i.e. the motion of the surface energy elements of the layer is geodesic. Since geodesic motion is a characteristic property of dust, we identify, in a final model-forming step, the mass-energy density ϵ with the baryonic mass-density μ_B,

$$\mu_B = \epsilon = \sigma_p e^{U-k} \delta(\zeta), \quad (1.119)$$

i.e. we interpret the disc limit as a dust limit (1.56), thus arriving at a disc of dust model, formally characterized by an energy-momentum tensor (1.113), (1.18) and the local energy-momentum balance $T^{ik}{}_{;k} = 0$, implying, again in a formal way, geodesic motion (1.54) and local baryonic mass conservation (1.55). Despite $\mu_B = \epsilon$, the baryonic mass M_0 as calculated from (1.58) differs from the energy-mass M in Equation (1.57), of course.

1.7.4 The mass-shedding limit

Our intuition tells us that if a star is rotating too quickly, its gravitational pull will no longer suffice to hold it together. If it is on the verge of losing mass, it is said to be rotating at the mass-shedding limit. The shedding of mass first sets in at the equator[17] and in Newtonian theory, this limit can be described by stipulating that the pressure gradient (more precisely, $\nabla p/\mu$) vanish there, thus implying that the gravitational force balances the centrifugal force in a corotating reference frame. In Einsteinian theory, $\nabla p/(\epsilon+p) \to 0$ implies that a fluid element follows a geodesic.

Focusing for the moment on Newtonian theory, we can consider the function

$$V = U - \frac{1}{2}\Omega^2 \varrho^2, \quad (1.120)$$

which is constant along the surface of the body:

$$V = V_0 \quad \text{along the boundary}, \quad (1.121)$$

see Subsection 1.7.1. We describe the surface by the parameterization $\zeta = \zeta_b(\varrho)$, restricting ourselves to the half-space $\zeta \geq 0$. Taking the derivative of (1.120) with respect to ϱ along the surface of the body yields

$$0 = \frac{\partial U}{\partial \varrho} + \frac{\partial U}{\partial \zeta}\frac{d\zeta_b}{d\varrho} - \Omega^2 \varrho. \quad (1.122)$$

Reflectional symmetry implies that

$$\left.\frac{\partial U}{\partial \zeta}\right|_{\zeta=0} = 0, \quad (1.123)$$

---

[17] In Newtonian theory, it can be proved that figures of equilibrium are always reflectionally symmetric with respect to the plane $\zeta = 0$ (Lichtenstein 1933). In Einsteinian theory, the same symmetry is to be expected, see Lindblom (1992), and we assume it to exist for this discussion.

from which it follows that at the equator

$$\frac{\partial U}{\partial \varrho} = \Omega^2 \varrho \quad \text{if} \quad \left|\frac{d\zeta_b}{d\varrho}\right| < \infty. \tag{1.124}$$

This equation tells us that the force due to gravity is equal in magnitude and opposite in direction to the centrifugal force. The inequality in (1.124) tells us that the surface does not meet the equatorial plane at a right angle. In other words, a cusp in the equatorial plane necessarily implies that the star is rotating at the mass-shedding limit. It can easily be verified that the same is true in Einsteinian theory. Moreover, numerical results suggest both in Newtonian and Einsteinian gravity that a cusp is a necessary and sufficient condition for the existence of a mass-shedding limit.

The potentials and surface function describing a mass-shedding star are not analytic, which makes a highly accurate description of them particularly challenging. For homogeneous Newtonian stars, it turns out that these functions are not even C^2. The proof of this as well as a more general discussion of differentiability can be found in Appendix A1.1. Despite the aforementioned challenges, the extremely simple Roche model for mass-shedding stars is very accurate in certain cases as is shown in Appendix A1.2.

1.8 Transition to black holes

1.8.1 Horizons

The event horizon of a stationary and axisymmetric black hole is given by a hypersurface $\mathcal{H}$ whose normal vector χ^i is a linear combination of the two Killing vectors ξ^i and η^i,

$$\chi^i \equiv \xi^i + \Omega_h \eta^i, \quad \Omega_h = \text{constant}, \tag{1.125}$$

and becomes null (lightlike) on that hypersurface:

$$\mathcal{H}: \quad \chi^i \chi_i = 0, \tag{1.126}$$

i.e. the horizon is a null hypersurface to which a Killing vector field is normal (a *Killing horizon*). Ω_h is called the 'angular velocity of the horizon', see Hawking and Ellis (1973) and Carter (1973). For a recent review of the status of the rigorous mathematical theory of stationary black holes, we refer the reader to Beig and Chruściel (2006). Because of the symmetries of the spacetime (and the horizon), each of the Killing vectors ξ^i and η^i must be tangential to the horizon, and therefore

$$\mathcal{H}: \quad \chi^i \xi_i = 0, \quad \chi^i \eta_i = 0. \tag{1.127}$$

For the metric in the form

$$ds^2 = e^{2\alpha}(d\varrho^2 + d\zeta^2) + W^2 e^{-2\nu}(d\varphi - \omega\,dt)^2 - e^{2\nu}dt^2, \tag{1.128}$$

the relations (1.125)–(1.127) lead to the following boundary conditions on the horizon, see Bardeen (1973a):

$$\mathcal{H}: \quad W = 0, \quad e^{2\nu} = 0, \quad \omega = \Omega_{\mathrm{h}}. \tag{1.129}$$

[Note that $W^2 = (\xi^i \eta_i)^2 - \xi^i \xi_i \eta^k \eta_k = (\chi^i \eta_i)^2 - \chi^i \chi_i \eta^k \eta_k$.] An immediate consequence of the condition $W = 0$ is the fact that in canonical Weyl coordinates, where $W \equiv \varrho$, the horizon is a part of the ζ-axis. This part must not be confused with the other parts of the ζ-axis, where $W = 0$ holds because the Killing vector η vanishes, defining the axis of symmetry.

1.8.2 Kerr black holes

Kerr black holes are the only stationary and axisymmetric, isolated black holes surrounded by a vacuum. This follows from the black hole uniqueness theorems, see Robinson (1975), Heusler (1996) and references therein, and Section 2.4, where the Kerr solution will be constructed as the unique solution to the black hole boundary value problem.

The Kerr metric (Kerr 1963) in Boyer–Lindquist coordinates (Boyer and Lindquist 1967) is given by

$$ds^2 = \Sigma\left(\frac{dr^2}{\Delta} + d\vartheta^2\right) + e^{-2\nu}\Delta \sin^2\vartheta\,(d\varphi - \omega\,dt)^2 - e^{2\nu}dt^2 \tag{1.130}$$

with

$$\Sigma = r^2 + (J/M)^2 \cos^2\vartheta, \quad \Delta = r^2 - 2Mr + (J/M)^2, \tag{1.131}$$

$$e^{2\nu} = \frac{\Delta \Sigma}{[r^2 + (J/M)^2]^2 - \Delta(J/M)^2 \sin^2\vartheta} \quad \text{and} \quad \omega = \frac{2Jr}{\Delta \Sigma}e^{2\nu}. \tag{1.132}$$

The Boyer–Lindquist coordinates r, ϑ are related to the canonical Weyl coordinates ϱ, ζ by

$$\varrho = \sqrt{r^2 - 2Mr + (J/M)^2}\,\sin\vartheta, \quad \zeta = (r - M)\cos\vartheta. \tag{1.133}$$

The Kerr solution depends on two parameters, the gravitational mass M and the angular momentum J (in this section we assume $J \geq 0$ without loss of generality).

The horizon of the black hole is given by

$$r = r_+ \equiv M + \sqrt{M^2 - (J/M)^2}, \tag{1.134}$$

the larger root of the quadratic equation $\Delta = 0$. Note that the Kerr metric describes a black hole only if

$$J \leq M^2 \tag{1.135}$$

is satisfied. The angular velocity of the horizon introduced in the previous subsection is given by

$$\Omega_h = \omega(r_+, \vartheta) = \text{constant} = \frac{J}{2M^2 \left[M + \sqrt{M^2 - (J/M)^2}\right]}, \tag{1.136}$$

cf. (1.129). The boundary of the ergosphere, see Subsection 1.6.2, is characterized by

$$r = r_0(\vartheta) \equiv M + \sqrt{M^2 - (J/M)^2 \cos^2 \vartheta}. \tag{1.137}$$

Within the ergosphere ($r_+ < r < r_0$), all observers must rotate in the same direction as the black hole ($d\varphi/dt > 0$), cf. (1.68).

It is interesting to discuss circular orbits of test particles in the equatorial 'plane' $\vartheta = \pi/2$. Their angular velocity is given by

$$\Omega = \pm \frac{\sqrt{M}}{r^{3/2} \pm J/\sqrt{M}}, \tag{1.138}$$

where the upper sign characterizes direct orbits (corotating with the black hole) and the lower sign holds for retrograde (counter-rotating) orbits. The circular orbits exist only for $r > r_{\text{ph}}$, with the 'photon orbit'

$$r_{\text{ph}} = 2M \left\{1 + \cos\left[\frac{2}{3}\arccos\left(\mp\frac{J}{M^2}\right)\right]\right\}. \tag{1.139}$$

The orbits are bound for $r > r_{\text{mb}}$, with the 'marginally bound orbit'

$$r_{\text{mb}} = 2M \mp J/M + 2M^{1/2}(M \mp J/M)^{1/2}. \tag{1.140}$$

A particle in an unbound orbit will escape to infinity under the influence of an infinitesimal outward perturbation. The orbits are stable for $r > r_{\text{ms}}$, with the 'marginally stable orbit'

$$r_{\text{ms}} = M \left\{3 + Z_2 \mp [(3 - Z_1)(3 + Z_1 + 2Z_2)]^{1/2}\right\}, \tag{1.141}$$

1.8 Transition to black holes

Table 1.1. *Photon orbit, marginally bound orbit and marginally stable orbit for the Schwarzschild black hole and for the extreme Kerr black hole. In the latter case one has to distinguish between direct and retrograde orbits.*

	r_{ph}	r_{mb}	r_{ms}
$J = 0$	$3M$	$4M$	$6M$
$J = M^2$ (direct)	M	M	M
$J = M^2$ (retrograde)	$4M$	$(3 + 2\sqrt{2})M$	$9M$

where

$$Z_1 = 1 + \left(1 - \frac{J^2}{M^4}\right)^{1/3} \left[\left(1 + \frac{J}{M^2}\right)^{1/3} + \left(1 - \frac{J}{M^2}\right)^{1/3}\right] \qquad (1.142)$$

and

$$Z_2 = \left(3\frac{J^2}{M^4} + Z_1^2\right)^{1/2}. \qquad (1.143)$$

These results on circular orbits of test particles were derived by Bardeen *et al.* (1972), see also Shapiro and Teukolsky (1983).

The two limiting cases of the Kerr black hole are $J = 0$, the non-rotating (Schwarzschild) black hole, and $J = M^2$, the maximally rotating (extreme Kerr) black hole. For $J = 0$, the horizon is given by $r_+ = 2M$ ('Schwarzschild radius') and no ergosphere exists. For $J = M^2$ one has $r_+ = M$ and the ergosphere, as with all $J > 0$, extends up to $r_0(\pi/2) = 2M$ in the equatorial plane. The values of the characteristic radii r_{ph}, r_{mb} and r_{ms} discussed above are given in Table 1.1. In the extreme case, $r_{ph}^{(d)}$, $r_{mb}^{(d)}$ and $r_{ms}^{(d)}$, where the (d) indicates direct orbits, all coincide with $r_+ = M$. However, defining proper radial distances ($\vartheta = $ constant, $\varphi = $ constant, $t = $ constant) according to

$$\delta(r_1, r_2) = \int_{r_1}^{r_2} \sqrt{g_{rr}}\, dr, \quad g_{rr} = \frac{\Sigma}{\Delta}, \qquad (1.144)$$

one finds for $\vartheta = \pi/2$ (Bardeen *et al.* 1972)

$$\lim_{J \to M^2} \delta(r_+, r_{ph}^{(d)}) = \frac{M}{2} \ln 3, \quad \lim_{J \to M^2} \delta(r_+, r_{mb}^{(d)}) = M \ln(1 + \sqrt{2}) \qquad (1.145)$$

and

$$\lim_{J \to M^2} \delta(r_+, r_{\text{ms}}^{(\text{d})}) = \infty. \qquad (1.146)$$

Moreover, because of the double zero of Δ at $r = M$ for the extreme Kerr metric, one obtains

$$\delta(M, r) = \infty \quad \text{for any } r > M \text{ and all } \vartheta. \qquad (1.147)$$

This means that the horizon (together with the whole '$r = M$ region') has an infinite proper radial distance from any point in the 'exterior' region $r > M$. On the other hand, the proper time needed for an infalling particle starting at some $r > M$ to reach the horizon remains finite. The geometrical situation can be illustrated nicely by embedding the $r \geq r_+$ part of the 'plane' $\vartheta = \pi/2$, $t = $ constant into three-dimensional Euclidean space (Bardeen et al. 1972), see Fig. 1.2. In the limit $J = M^2$, an infinitely long 'throat' characterized by $r = M$ (circumference: $4\pi M$) appears. The horizon is situated at the bottom and the direct orbits corresponding to $r_{\text{ph}}^{(\text{d})}$, $r_{\text{mb}}^{(\text{d})}$ and $r_{\text{ms}}^{(\text{d})}$ are located at different places along the throat. However, the proper time of an infalling particle needed to pass through this throat is zero. Bardeen and Horowitz (1999) have studied the 'throat geometry' ($r = M$) by means of the coordinate transformation

$$\tilde{r} = \frac{r - M}{\lambda}, \quad \tilde{\vartheta} = \vartheta, \quad \tilde{\varphi} = \varphi - \Omega_{\text{h}} t, \quad \tilde{t} = \lambda t \qquad (1.148)$$

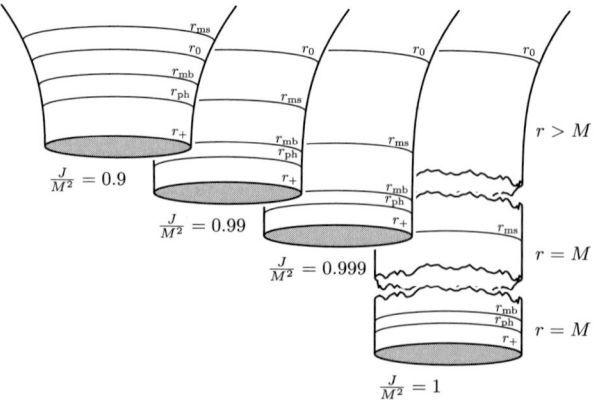

Fig. 1.2. Embedding diagrams of the $r \geq r_+$ part of the 'plane' $\vartheta = \pi/2$, $t = $ constant of the Kerr metric as it approaches the limit $J = M^2$. This 2-surface is represented here as a surface of revolution in three-dimensional Euclidean space with the same interior geometry. The positions of the direct orbits r_{ph}, r_{mb}, r_{ms} and of the boundary $r_0(\pi/2)$ of the ergosphere are shown. Note that the two gaps in the fourth picture are both infinitely long (adapted from Bardeen et al. 1972).

1.8 Transition to black holes

in the limit $\lambda \to 0$. According to (1.136), Ω_h is given here by

$$\Omega_h = \frac{1}{2M} \qquad (J = M^2). \tag{1.149}$$

One obtains

$$ds^2 = M^2(1 + \cos^2 \tilde{\vartheta}) \left(\frac{d\tilde{r}^2}{\tilde{r}^2} + d\tilde{\vartheta}^2 - \frac{\tilde{r}^2 d\tilde{t}^2}{4M^4} \right) + \frac{4M^2 \sin^2 \tilde{\vartheta}}{1 + \cos^2 \tilde{\vartheta}} \left(d\tilde{\varphi} + \frac{\tilde{r} d\tilde{t}}{2M^2} \right)^2. \tag{1.150}$$

This represents a completely non-singular vacuum solution of Einstein's equations, which is geodesically complete but no longer asymptotically flat, see Bardeen and Horowitz (1999). The area of all surfaces $\tilde{r} = $ constant, $\tilde{t} = $ constant is $8\pi M^2$ and equal to the area of the horizon of the extreme Kerr black hole. In addition to $\partial/\partial\tilde{t}$ and $\partial/\partial\tilde{\varphi}$, the metric (1.150) has two more Killing fields (Wolf 1998):

$$\left(\frac{\tilde{t}^2}{2} + \frac{1}{8\tilde{r}^2 \Omega_h^4} \right) \frac{\partial}{\partial \tilde{t}} - \tilde{r}\tilde{t} \frac{\partial}{\partial \tilde{r}} - \frac{1}{2\tilde{r} \Omega_h^2} \frac{\partial}{\partial \tilde{\varphi}}, \quad \tilde{t} \frac{\partial}{\partial \tilde{t}} - \tilde{r} \frac{\partial}{\partial \tilde{r}}. \tag{1.151}$$

1.8.3 From rotating fluids to black holes

The condition characterizing a black hole horizon,

$$(\xi^i + \Omega_h \eta^i)(\xi_i + \Omega_h \eta_i) = 0, \tag{1.152}$$

looks similar to the surface condition of a rotating fluid body,

$$(\xi^i + \Omega \eta^i)(\xi_i + \Omega \eta_i) = -e^{2V_0}. \tag{1.153}$$

An interesting question is whether or not there exist continuous parametric transitions from stationary fluid configurations to black holes. We now want to show that

$$V_0 \to -\infty \tag{1.154}$$

or, equivalently,

$$M - 2\Omega J \to 0 \tag{1.155}$$

is necessary and sufficient for a black hole limit of a fluid body in equilibrium (Meinel 2004, 2006). A remarkable consequence of (1.155) is the fact that such a limit always leads to the extreme Kerr black hole.

In a first step, we prove the equivalence of (1.154) and (1.155). To this end, we combine formulae (1.57) and (1.58) for the gravitational mass M, the angular momentum J and the baryonic mass M_0, leading to

$$M = 2\Omega J + \int \frac{\epsilon + 3p}{\mu_B} e^V dM_0, \tag{1.156}$$

where we have used Equations (1.18), (1.21) and the abbreviation $dM_0 = -\mu_B u_i n^i dV$ for the baryonic mass elements. With (1.23) and (1.26), we get

$$M = 2\Omega J + h(0) e^{V_0} \int \frac{\epsilon + 3p}{\epsilon + p} dM_0. \tag{1.157}$$

Since, for non-negative ϵ and p, $1 \leq (\epsilon + 3p)/(\epsilon + p) \leq 3$ holds, conditions (1.154) and (1.155) are equivalent.[18]

Next we note that all linear combinations of the two Killing vectors ξ^i and η^i that are not proportional to the null (lightlike) vector $\xi^i + \Omega_h \eta^i$ become spacelike on the horizon. However, the 4-velocity (1.18) must always be timelike. Therefore, a black hole limit of a fluid body can only be approached for

$$\Omega \to \Omega_h \tag{1.158}$$

together with $V_0 \to -\infty$, i.e. (1.154) is necessary.

Finally, we prove that (1.154) is also sufficient for a black hole limit. Because of (1.19) and (1.26), the surface of the fluid is characterized in general by

$$\chi^i \chi_i = -e^{2V_0}, \quad \chi^i \equiv \xi^i + \Omega \eta^i. \tag{1.159}$$

The Killing vector χ^i is tangential to the hypersurface $\mathcal{H}$ generated by the timelike world lines of the fluid elements of the surface of the body with 4-velocity $u^i = e^{-V_0} \chi^i$, see (1.18). Each of the Killing vectors ξ^i and η^i must itself be tangential to $\mathcal{H}$ because of the symmetries of the spacetime. In the limit $V_0 \to -\infty$, we approach a situation in which χ^i becomes null on $\mathcal{H}$:

$$\chi^i \chi_i \to 0. \tag{1.160}$$

Moreover, with the reasonable assumption[19]

$$0 \leq -\xi^i u_i \leq 1, \tag{1.161}$$

---

[18] We assume $0 < M_0 < \infty$ and $0 < h(0) < \infty$.

[19] The condition $-\xi^i u_i \leq 1$ ensures that a particle resting on the surface of the fluid is (at least marginally) bound, i.e. cannot escape to infinity on a geodesic; $-\xi^i u_i \geq 0$ follows from $-\xi^i u_i = -(\chi^i - \Omega \eta^i) u_i = e^{V_0} + \Omega \eta^i u_i$, since $\eta^i u_i$ will always have the same sign as Ω ($\eta^i u_i = 0$ on the axis, of course).

1.8 Transition to black holes

we find that χ^i also becomes orthogonal to ξ^i (and thus to η^i) on $\mathcal{H}$ in the limit:

$$\chi^i \xi_i \to 0, \quad \chi^i \eta_i \to 0. \tag{1.162}$$

Together with the orthogonal transitivity of the spacetime (see Section 1.3), χ^i therefore becomes orthogonal to three linearly independent tangent vectors at each point of $\mathcal{H}$, i.e. normal to $\mathcal{H}$. Because of (1.160), we thus approach a situation in which $\mathcal{H}$ is a null hypersurface and satisfies all defining conditions for a horizon of a stationary and axisymmetric black hole with Ω being the angular velocity of the horizon, see Subsection 1.8.1. According to the black hole uniqueness theorems, we conclude that (outside the horizon) the Kerr metric with $J \leq M^2$ results.[20] Then, with (1.155), (1.158) and (1.136), we are necessarily led to the case $J = M^2$. Therefore, the metric of an extreme Kerr black hole (outside the horizon) results, whenever a sequence of fluid bodies admits a limit $V_0 \to -\infty$.

Later in this book, we shall demonstrate with analytical as well as numerical examples that such parametric ('quasi-stationary') transitions from fluid bodies to black holes are indeed possible. It will turn out that an interesting 'separation of spacetimes' with a non-asymptotically flat 'inner world' (containing the fluid body and its surroundings) and the $r > M$ part of the extreme Kerr metric as the 'outer world' emerges in the limit $V_0 \to -\infty$. The behaviour of the 'inner world' at spatial infinity is given precisely by the 'throat geometry' (1.150).

---

[20] Note that the black hole uniqueness proof by construction given in Neugebauer and Meinel (2003) can be extended to the case in which the horizon is degenerate, leading to $J = M^2$. This will be shown in Section 2.4.

2
Analytical treatment of limiting cases

A particular difficulty of the rotating fluid body problem – in Newton's as well as in Einstein's theory – is the free boundary of the fluid, which is not known from the outset. However, in some special cases, the shape of the surface is known (i) by making a lucky guess, (ii) from symmetry considerations leading to spheres in the non-rotating case, and (iii) by considering the limit of extreme flattening leading to infinitesimally thin discs.

2.1 Maclaurin spheroids

An important example for rotating figures of equilibrium within Newton's theory of gravity – the Maclaurin spheroids – was indeed found by a lucky guess for the surface's shape. It turns out that homogeneous (constant mass-density) spheroids are solutions to the problem for a certain constant angular velocity, given prescribed values of the mass density and the semi-axes of the spheroid.

The Newtonian potential can of course be calculated from the Poisson integral. For ellipsoidal configurations the choice of elliptic coordinates proves useful. The exterior potential of a homogeneous oblate spheroid of mass M and focal length ϱ_0 is given by

$$U = -\frac{M}{\varrho_0}\left\{\operatorname{arccot}\xi + \frac{3}{4}\left[\xi - \left(\xi^2 + \frac{1}{3}\right)\operatorname{arccot}\xi\right](1-3\eta^2)\right\}, \qquad (2.1)$$

with oblate elliptic coordinates ξ, η (and φ) related to the cylindrical coordinates ϱ, ζ (and φ) by

$$\varrho = \varrho_0\sqrt{(1+\xi^2)(1-\eta^2)}, \quad \zeta = \varrho_0\xi\eta \quad (0 \le \xi < \infty, -1 \le \eta \le 1), \qquad (2.2)$$

see Fig. 2.1. U can be written in the form

$$U = A(\xi) + B(\xi)\varrho^2, \qquad (2.3)$$

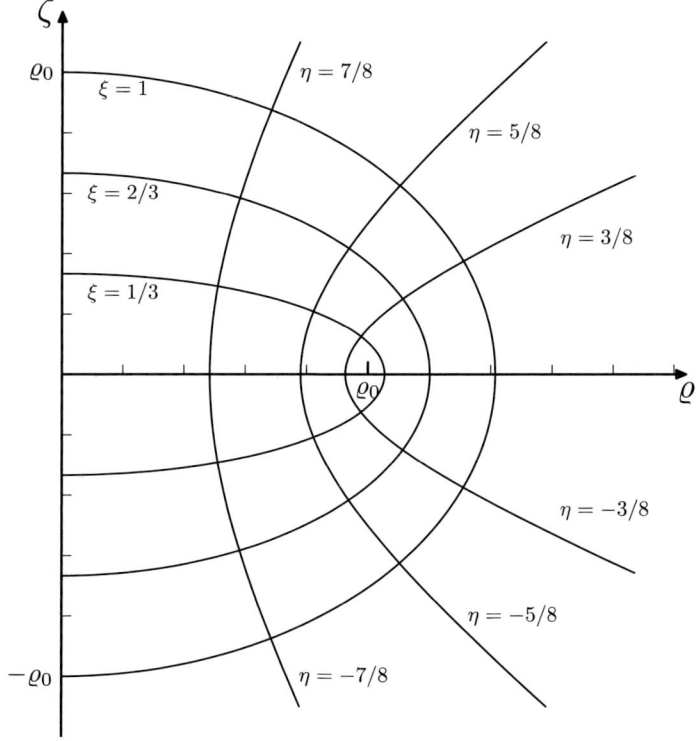

Fig. 2.1. A depiction of the oblate elliptic coordinates of Equation (2.2). Lines of constant ξ are ellipses and lines of constant η are hyperbolas all with the focus $(\varrho_0, 0)$.

with
$$A(\xi) = -\frac{3M}{2\varrho_0}\left[(\xi^2+1)\operatorname{arccot}\xi - \xi\right] \tag{2.4}$$
and
$$B(\xi) = \frac{3M\left[(3\xi^2+1)\operatorname{arccot}\xi - 3\xi\right]}{4\varrho_0^3(1+\xi^2)}. \tag{2.5}$$

Let the surface of the spheroid be characterized by $\xi = \xi_0$. Then the surface condition
$$V \equiv U - \frac{1}{2}\Omega^2\varrho^2 = V_0 = \text{constant} \quad (\xi = \xi_0) \tag{2.6}$$
is satisfied for
$$\Omega = \sqrt{2B(\xi_0)}, \quad V_0 = A(\xi_0). \tag{2.7}$$

Note that the semi-axes ($a = b > c$) of the ellipsoid are
$$a = \varrho_0\sqrt{1+\xi_0^2}, \quad c = \varrho_0\xi_0. \tag{2.8}$$

The mass is related to the constant mass-density μ by

$$M = \frac{4\pi}{3}\mu a^2 c. \qquad (2.9)$$

The interior solution ($\xi < \xi_0$) is given by

$$U = V_0 + \frac{1}{2}\Omega^2 \varrho^2 - C\left(1 - \frac{\varrho^2}{a^2} - \frac{\zeta^2}{c^2}\right), \qquad (2.10)$$

with

$$C = \frac{3M}{2\varrho_0}\xi_0(1 - \xi_0 \operatorname{arccot} \xi_0). \qquad (2.11)$$

The pressure distribution follows from the integrated Euler equation:

$$p = \mu(V_0 - V) = \mu C\left(1 - \frac{\varrho^2}{a^2} - \frac{\zeta^2}{c^2}\right). \qquad (2.12)$$

It can easily be verified that expressions (2.1) and (2.10) satisfy the Laplace and Poisson equations $\nabla^2 U = 0$ and $\nabla^2 U = 4\pi\mu$ respectively. The continuity of the potential at $\xi = \xi_0$ is obvious; the continuity of its normal derivative can easily be checked as well.

For a given value of the mass-density μ (i.e. for a given equation of state) the complete solution depends on two parameters, for example a and c or ϱ_0 and ξ_0. It is, of course, regular everywhere and has the correct asymptotic behaviour

$$U \to -\frac{M}{r} \quad \text{as} \quad r \to \infty \quad (r^2 \equiv \varrho^2 + \zeta^2). \qquad (2.13)$$

In the following, we present some further useful parameter relations. The angular velocity given by (2.7) can also be related to the mass-density and the eccentricity

$$\epsilon = \sqrt{1 - \frac{c^2}{a^2}} = \frac{1}{\sqrt{1 + \xi_0^2}} \qquad (2.14)$$

leading to the celebrated formula (Maclaurin 1742)

$$\frac{\Omega^2}{\pi\mu} = \frac{2}{\epsilon^3}\sqrt{1 - \epsilon^2}(3 - 2\epsilon^2)\arcsin\epsilon - \frac{6}{\epsilon^2}(1 - \epsilon^2), \qquad (2.15)$$

a plot of which can be found in Fig. 2.2. The angular momentum J and kinetic energy of rotation E_{rot} are given by

$$J = I\Omega, \quad E_{\text{rot}} = \frac{1}{2}I\Omega^2, \qquad (2.16)$$

2.1 Maclaurin spheroids

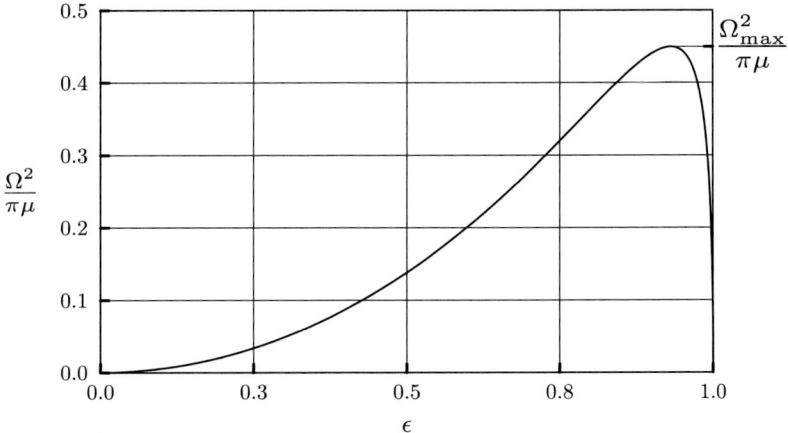

Fig. 2.2. The normalized, squared angular velocity $\Omega^2/\pi\mu$ for the Maclaurin sequence as it depends on eccentricity ϵ, see Equation (2.15). For a given mass-density μ and for some $\Omega < \Omega_{\max}$, there exist solutions for two different values of ϵ.

with the moment of inertia about the axis of rotation

$$I = \frac{2}{5}Ma^2. \qquad (2.17)$$

The gravitational (potential) energy is

$$E_g \equiv \frac{1}{2}\int U\mu\, d^3x = -\frac{3M^2 \arcsin \epsilon}{5a\epsilon}. \qquad (2.18)$$

It is interesting to discuss the two limiting cases of the Maclaurin spheroids:

(i) Maclaurin spheres

For $\epsilon \to 0$ we get a (non-rotating) sphere of radius $R \equiv a = c$ with

$$U = \begin{cases} -2\pi\mu(R^2 - r^2/3): & r < R \\ -M/r: & r > R \end{cases} \qquad (2.19)$$

and

$$p = \frac{2\pi\mu^2}{3}(R^2 - r^2) \qquad (r \le R). \qquad (2.20)$$

The gravitational energy of a homogeneous sphere is

$$E_g = -\frac{3M^2}{5R}. \qquad (2.21)$$

(ii) Maclaurin discs

For $\epsilon \to 1$ we get a disc with radius ϱ_0. The potential U is given by (2.1) everywhere. On the disc itself (characterized by $\xi = 0$ or, equivalently, $\zeta = 0$

and $\varrho \leq \varrho_0$)
$$U = V_0 + \frac{1}{2}\Omega^2\varrho^2, \qquad (2.22)$$
i.e. $V = V_0 =$ constant, results. Equation (2.9) shows that $\mu \to \infty$ in the disc limit.[1] The corresponding *surface mass-density* σ is a function of ϱ:
$$\sigma = \frac{3M}{2\pi\varrho_0^3}\sqrt{\varrho_0^2 - \varrho^2}. \qquad (2.23)$$

It can be calculated either by simple geometric means or via the jump in the normal derivative of U: $U_{,\zeta}|_{\zeta=0^+} - U_{,\zeta}|_{\zeta=0^-} = 2U_{,\zeta}|_{\zeta=0^+} = 4\pi\sigma$. Note that $p/\mu \to 0$ holds in the disc limit, since the central (maximal) pressure remains finite. Accordingly, the solution may also be interpreted as a 'rigidly rotating disc of dust', see Subsection 1.7.3. This corresponds to the fact that for each (surface-) mass element the gravitational force and the centrifugal force are in equilibrium ($\partial U/\partial\varrho = \Omega^2\varrho$). Angular velocity, mass and radius are related according to
$$\Omega^2 = \frac{3\pi M}{4\varrho_0^3} \qquad (2.24)$$
and the relation
$$V_0 = -\Omega^2\varrho_0^2 \qquad (2.25)$$
holds. The angular momentum is
$$J = \frac{8}{15\pi}\Omega^3\varrho_0^5 \qquad (2.26)$$
and the kinetic and gravitational energies are given by
$$E_{\text{rot}} = \frac{3\pi M^2}{20\varrho_0} \quad \text{and} \quad E_g = -\frac{3\pi M^2}{10\varrho_0}. \qquad (2.27)$$

Remarkably, analytic solutions of the Einstein equations exist for both limiting cases (spheres and discs) as well. These will be treated in the next two sections.

2.2 Schwarzschild spheres

The global solution of Einstein's field equations describing a perfect fluid sphere of constant mass(-energy) density μ was found by Schwarzschild (1916). In our coordinates, the line element is
$$ds^2 = e^{2\alpha}(d\varrho^2 + d\zeta^2 + \varrho^2 d\varphi^2) - e^{2\nu}dt^2, \qquad (2.28)$$

---
[1] We assume finite values for the disc's radius and mass, of course.

2.2 Schwarzschild spheres

where α and ν depend on

$$r = \sqrt{\varrho^2 + \zeta^2} \qquad (2.29)$$

only. We denote the coordinate radius of the sphere by r_0. Inside the fluid ($r < r_0$) we have the 'interior Schwarzschild solution' given by

$$e^{\alpha} = \frac{\left(1 + \frac{M}{2r_0}\right)^3}{1 + \frac{Mr^2}{2r_0^3}}, \quad e^{\nu} = \frac{1}{2}\left(3\frac{1 - \frac{M}{2r_0}}{1 + \frac{M}{2r_0}} - \frac{1 - \frac{Mr^2}{2r_0^3}}{1 + \frac{Mr^2}{2r_0^3}}\right) \qquad (2.30)$$

and

$$p = \mu(e^{\nu_0 - \nu} - 1), \quad e^{\nu_0} \equiv e^{\nu(r_0)}. \qquad (2.31)$$

The exterior solution ($r > r_0$) is naturally the vacuum Schwarzschild metric given by

$$e^{\alpha} = \left(1 + \frac{M}{2r}\right)^2, \quad e^{\nu} = \frac{1 - \frac{M}{2r}}{1 + \frac{M}{2r}}. \qquad (2.32)$$

Of course, in the ('isotropic') coordinates used here, the metric coefficients and their first derivatives are continuous at the surface of the fluid.[2] M is the gravitational mass. For a given value of the mass-density μ, the solution depends on only one parameter, say ν_0. M and r_0 are related to ν_0 according to

$$M = \frac{1}{4}\sqrt{\frac{3}{2\pi\mu}}\left(1 - e^{2\nu_0}\right)^{3/2}, \qquad (2.33)$$

$$\frac{M}{2r_0} = \frac{1 - e^{\nu_0}}{1 + e^{\nu_0}}. \qquad (2.34)$$

It turns out that

$$e^{\nu_0} > \frac{1}{3} \qquad (2.35)$$

must hold. For $e^{\nu_0} \to 1/3$, the central pressure tends to infinity.[3] As a consequence, the relative redshift

$$z = e^{-\nu_0} - 1 \qquad (2.36)$$

of photons emitted from the fluid's surface and received at infinity is bounded by the value 2. Even more importantly, the gravitational mass is also bounded:

$$M < M_{\max} = \frac{4}{9\sqrt{3\pi\mu}}. \qquad (2.37)$$

---

[2] The more familiar radial Schwarzschild coordinate $\tilde{r}$ is related to our r by $\tilde{r} = r(1 + M/2r)^2$ in the vacuum region and $\tilde{r} = 2Br/(B^2 + Ar^2)$ with $A = 8\pi\mu/3$, $B = \frac{1}{4}(1 + e^{\nu_0})^3$ inside the fluid. In terms of the Schwarzschild coordinate radius $\tilde{r}_0$, the gravitational mass is given by the simple expression $M = \frac{4\pi}{3}\mu\tilde{r}_0^3$. Note that $\partial g_{\tilde{r}\tilde{r}}/\partial\tilde{r}$ is discontinuous at $\tilde{r} = \tilde{r}_0$.
[3] In this limit, the Schwarzschild coordinate radius $\tilde{r}_0$ reaches the 'Buchdahl limit' $9M/4$, see Subsection 1.7.2.

The baryonic mass (calculated under the assumption $\mu = \mu_B$, cf. Section 1.5) is

$$M_0 = \frac{3}{8}\sqrt{\frac{3}{2\pi\mu}}\left(\arccos e^{v_0} - e^{v_0}\sqrt{1 - e^{2v_0}}\right). \tag{2.38}$$

The relative binding energy $(M_0 - M)/M_0$ is positive and reaches a maximal value of about 0.39 in the limit $e^{v_0} \to 1/3$.

The Newtonian limit of the Schwarzschild spheres is approached for $\gamma \equiv 1 - e^{v_0} \ll 1$ and leads back to the Maclaurin spheres. Of course, $M/M_0 \to 1$ as $\gamma \to 0$. However, it is interesting to note that the leading order term in an expansion of $M - M_0$ in powers of $\gamma^{1/2}$ gives exactly the Newtonian gravitational energy E_g, see (2.21).

2.3 The rigidly rotating disc of dust

2.3.1 The boundary value problem of the Ernst equation

As has already been shown in Subsection 1.7.3, the model describing a uniformly rotating, infinitely thin disc leads to a boundary value problem for the Ernst equation, see Fig. 2.3. On the disc, i.e. for $\zeta = 0$ and $\varrho \leq \varrho_0$, the Ernst potential f' in the corotating frame has to be equal to a real constant $\exp(2V_0)$ with $V_0 < 0$,

$$f' = e^{2V_0} \quad \text{for} \quad \zeta = 0, \varrho \leq \varrho_0, \tag{2.39}$$

see (1.107), whereas the Ernst potential f in the non-rotating frame has to be regular everywhere outside the disc and has to satisfy

$$f \to 1 \quad \text{as} \quad \varrho^2 + \zeta^2 \to \infty, \tag{2.40}$$

see (1.108).

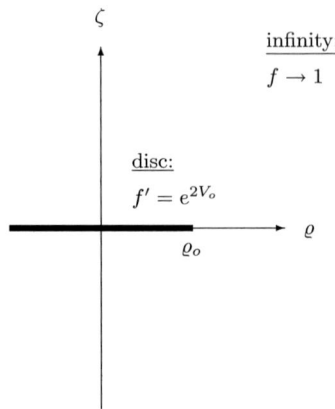

Fig. 2.3. The boundary value problem of the Ernst equation for a rigidly rotating disc of dust.

At first glance, the boundary value problem seems to depend on the three parameters V_0, ϱ_0 and Ω, where the angular velocity Ω enters into the problem via the relation between f and f'. However, it will turn out that the solution is not regular at the rim of the disc unless a certain relation between the three parameters is satisfied.[4] Consequently, only two of them can be chosen arbitrarily.

In the next subsections we shall derive the solution to this boundary value problem (Neugebauer and Meinel 1993, 1994, 1995, Neugebauer *et al.* 1996, Neugebauer and Meinel 2003). This will be done by applying the 'inverse method',[5] which relies on the existence of a 'linear problem' for the Ernst equation (Belinski and Zakharov 1978, Harrison 1978, Maison 1978, Hoenselaers *et al.* 1979, Neugebauer 1979, 1980a). A further application of this method, namely the derivation of the Kerr metric as the unique solution for a stationary and axisymmetric vacuum black hole, including the degenerate case, will then be given in Section 2.4.

2.3.2 Solution via the 'inverse method'

Neugebauer's form of the linear problem (LP) for the Ernst equation reads (Neugebauer 1980b, Neugebauer and Kramer 1983)

$$\Phi_{,z} = \left\{ \begin{pmatrix} B & 0 \\ 0 & A \end{pmatrix} + \lambda \begin{pmatrix} 0 & B \\ A & 0 \end{pmatrix} \right\} \Phi,$$

$$\Phi_{,\bar{z}} = \left\{ \begin{pmatrix} \bar{A} & 0 \\ 0 & \bar{B} \end{pmatrix} + \frac{1}{\lambda} \begin{pmatrix} 0 & \bar{A} \\ \bar{B} & 0 \end{pmatrix} \right\} \Phi, \tag{2.41}$$

where $\Phi(z, \bar{z}, \lambda)$ is a 2×2 matrix depending on the spectral parameter

$$\lambda = \sqrt{\frac{K - i\bar{z}}{K + iz}} \tag{2.42}$$

as well as on the complex coordinates

$$z = \varrho + i\zeta, \quad \bar{z} = \varrho - i\zeta, \tag{2.43}$$

whereas A, B and their complex conjugates $\bar{A}, \bar{B}$ are functions of $z, \bar{z}$ (or ϱ, ζ) and do not depend on K. From the integrability condition and the identities

$$\lambda_{,z} = \frac{\lambda}{4\varrho}\left(\lambda^2 - 1\right), \quad \lambda_{,\bar{z}} = \frac{1}{4\varrho\lambda}\left(\lambda^2 - 1\right), \tag{2.44}$$

---

[4] In the Newtonian limit ($|V_0| \ll 1$) this parameter relation reduces to (2.25).
[5] For a general introduction to the 'inverse method' in the context of soliton theory, see e.g. Novikov *et al.* (1984).

it follows that a certain matrix polynomial in λ has to vanish. This yields the set of first order differential equations

$$A_{,\bar{z}} = A(\bar{B} - \bar{A}) - \frac{1}{4\varrho}(A + \bar{B}), \quad B_{,\bar{z}} = B(\bar{A} - \bar{B}) - \frac{1}{4\varrho}(B + \bar{A}). \tag{2.45}$$

The system has the 'first integrals'

$$A = \frac{f_{,z}}{f + \bar{f}}, \quad B = \frac{\bar{f}_{,z}}{f + \bar{f}}. \tag{2.46}$$

Resubstituting A and B in Equations (2.45), one obtains the Ernst equation

$$(\Re f)\left(f_{,\varrho\varrho} + f_{,\zeta\zeta} + \frac{1}{\varrho}f_{,\varrho}\right) = f_{,\varrho}^2 + f_{,\zeta}^2, \tag{2.47}$$

cf. (1.43). Thus, the Ernst equation is the integrability condition of the LP (2.41). Conversely, if f is a solution to the Ernst equation, the matrix Φ calculated from (2.41) does not depend on the path of integration. The idea of the 'inverse method' is to discuss Φ for fixed, but arbitrary, values of z and $\bar{z}$ as a holomorphic function of λ (or K) and to calculate A, B and finally f afterwards. To obtain the desired information about the holomorphic structure in λ, we shall integrate the LP along the dotted line in Fig. 2.4. $\mathcal{A}^{\pm}$ are the two parts of the axis of symmetry ($\varrho = 0, \zeta > 0$ or $\zeta < 0$), $\mathcal{B}$ represents the surface of the disc ($\varrho \leq \varrho_0, \zeta = 0^{\pm}$) and $\mathcal{C}$ stands for spatial infinity.

In this way, we shall solve the *direct problem* of the inverse method and obtain $\Phi(z, \bar{z}, \lambda)$ for $z, \bar{z} \in \mathcal{A}^{\pm}, \mathcal{B}, \mathcal{C}$. It turns out that the holomorphic structure remains unchanged by an extension to values of z and $\bar{z}$ off the axis of symmetry into the

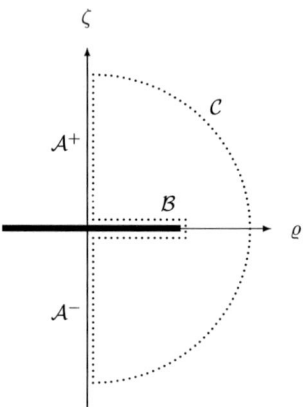

Fig. 2.4. The line of integration for the direct problem, hugging the disc, along the axis of symmetry and at infinity.

2.3 The rigidly rotating disc of dust

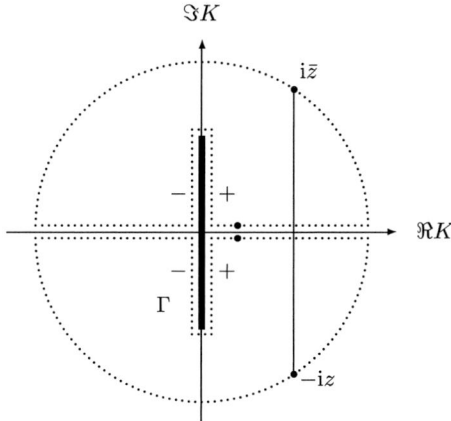

Fig. 2.5. The two-sheeted Riemann K-surface with moving branch cut (shown for some arbitrary z belonging to the path $\mathcal{A}^+\mathcal{C}\mathcal{A}^-\mathcal{B}$ of Fig. 2.4) and the 'image' of the disc, i.e. the contour(s) $\Gamma : \Re K = 0, -\varrho_0 \leq \Im K \leq \varrho_0$ (thick line).

entire vacuum region, meaning that one can construct functions Φ with prescribed properties in λ from which one obtains the desired solution $f(z, \bar{z})$ everywhere in the vacuum region. In some circumstances, λ may be replaced by K. For this purpose, it may be helpful to discuss the mapping (2.42) of the two-sheeted Riemann surface of K onto the λ-plane for different values of ϱ, ζ (or equivalently $z, \bar{z}$).

Figure 2.5 shows the position of the branch points $K_B = i\bar{z}, \bar{K}_B = -iz$ for the marked path $\mathcal{A}^+\mathcal{C}\mathcal{A}^-\mathcal{B}$ of Fig. 2.4 by two dotted curves. It reflects the slice $\varphi = \text{constant}, t = \text{constant}$ (Fig. 2.4) and indicates, in particular, the position of the disc.

Consider now a Riemann surface with confluent branch points $K_B = \bar{K}_B = \zeta \in \mathcal{A}^+$. Here λ degenerates and takes the values $\lambda = -1$ for K's in the lower sheet, say, and $\lambda = +1$ for K's in the upper sheet ($K \neq K_B$).

We shall now travel along the dotted line of Fig. 2.4 starting from and returning to any point $\varrho = 0, \zeta \in \mathcal{A}^+$. In Fig. 2.5 this corresponds to the bold faced points on the real axis. Note that $\lambda = -1$ for all K's ($K \neq \zeta$) in the lower and $\lambda = +1$ for all K's ($K \neq \zeta$) in the upper sheet of the Riemann K-surface belonging to axis values $\varrho = 0, \zeta \in \mathcal{A}^\pm$. The corresponding branch points cling to either side of the real axis in Fig. 2.5. For $\varrho, \zeta \in \mathcal{C}$, the cut between the branch points (e.g. the right solid line in Fig. 2.5) sweeps over the entire K-surface and puts 'upper' K values into the lower sheet and 'lower' K values into the 'upper' sheet. As a consequence, λ will change from ± 1 to ∓ 1 between $\varrho = 0, \zeta = +\infty$ and $\varrho = 0, \zeta = -\infty$. This 'exchange of sheets' is important for the solution of the LP: The initial value $\Phi(\varrho_0, \zeta_0, \lambda)$ can (and must) be fixed in only *one* sheet of the K-surface. The dependence on K in the other sheet follows by integrating the LP (2.41) along a suitable path.

We shall divide the integration of the LP (2.41) along the closed dotted line of Fig. 2.4 into two steps:

(i) Integrating along $\mathcal{A}^+\mathcal{C}\mathcal{A}^-$.
    This step leads to a '*general* solution' for $\boldsymbol{\Phi}$ on the regular parts $\mathcal{A}^\pm$ of the symmetry axis and will be used in Section 2.4 as well.
(ii) Integrating along $\mathcal{B}$.

Axis and infinity

It can easily be verified that the matrix coefficients $\mathbf{U}$ and $\mathbf{V}$ of the LP $\boldsymbol{\Phi}_{,z} = \mathbf{U}\boldsymbol{\Phi}$, $\boldsymbol{\Phi}_{,\bar{z}} = \mathbf{V}\boldsymbol{\Phi}$, as given in (2.41), satisfy the relations

$$\mathbf{U}(-\lambda) = \begin{pmatrix} 1 & 0 \\ 0 & -1 \end{pmatrix} \mathbf{U}(\lambda) \begin{pmatrix} 1 & 0 \\ 0 & -1 \end{pmatrix} \qquad (2.48)$$

and

$$\overline{\mathbf{U}\left(\frac{1}{\bar{\lambda}}\right)} = \begin{pmatrix} 0 & 1 \\ 1 & 0 \end{pmatrix} \mathbf{U}(\lambda) \begin{pmatrix} 0 & 1 \\ 1 & 0 \end{pmatrix}. \qquad (2.49)$$

Therefore, without loss of generality, the matrix $\boldsymbol{\Phi}$ may be assumed to have the structure

$$\boldsymbol{\Phi} = \begin{pmatrix} \psi(\varrho,\zeta,\lambda) & \psi(\varrho,\zeta,-\lambda) \\ \chi(\varrho,\zeta,\lambda) & -\chi(\varrho,\zeta,-\lambda) \end{pmatrix} \qquad (2.50)$$

together with

$$\overline{\psi\left(\varrho,\zeta,\frac{1}{\bar{\lambda}}\right)} = \chi(\varrho,\zeta,\lambda). \qquad (2.51)$$

The particular form of (2.50) is equivalent to

$$\boldsymbol{\Phi}(-\lambda) = \begin{pmatrix} 1 & 0 \\ 0 & -1 \end{pmatrix} \boldsymbol{\Phi}(\lambda) \begin{pmatrix} 0 & 1 \\ 1 & 0 \end{pmatrix} \qquad (2.52)$$

and (2.51) can be written as

$$\overline{\boldsymbol{\Phi}\left(\frac{1}{\bar{\lambda}}\right)} = \begin{pmatrix} 0 & 1 \\ 1 & 0 \end{pmatrix} \boldsymbol{\Phi}(\lambda) \begin{pmatrix} 1 & 0 \\ 0 & -1 \end{pmatrix}. \qquad (2.53)$$

For $K \to \infty$ and $\lambda = -1$, the functions ψ, χ may be normalized by

$$\psi(\varrho,\zeta,-1) = \chi(\varrho,\zeta,-1) = 1. \qquad (2.54)$$

Finally, the solution to the Ernst equation can be read off at $\lambda = 1$ ($K \to \infty$),

$$f(\varrho,\zeta) = \chi(\varrho,\zeta,1), \quad \overline{f(\varrho,\zeta)} = \psi(\varrho,\zeta,1), \qquad (2.55)$$

cf. (2.41) for $\lambda = 1$ and (2.46).

As discussed in Section 1.4, the Ernst equation retains its form in the corotating frame of reference. This ensures the existence of an LP (2.41) in the corotating system. The Φ-matrices of the two systems of reference are connected by the relation

$$\Phi' = \left[\begin{pmatrix} 1 + \Omega a - \Omega\varrho e^{-2U} & 0 \\ 0 & 1 + \Omega a + \Omega\varrho e^{-2U} \end{pmatrix} + i(K + iz)\Omega e^{-2U} \begin{pmatrix} -1 & -\lambda \\ \lambda & 1 \end{pmatrix} \right] \Phi. \tag{2.56}$$

As in Section 1.3, a prime marks 'corotating' quantities.

We are now in a position to integrate the LP (2.41) along the part $\mathcal{A}^+\mathcal{C}\mathcal{A}^-$ of the dotted line in Fig. 2.4. Using (2.41) along $\mathcal{A}^{\pm}$ and (2.46) one finds for the axis values of Φ

$$\mathcal{A}^+: \quad \Phi = \begin{pmatrix} \overline{f(\zeta)} & 1 \\ f(\zeta) & -1 \end{pmatrix} \begin{pmatrix} F(K) & 0 \\ G(K) & 1 \end{pmatrix}, \tag{2.57}$$

$$\mathcal{A}^-: \quad \Phi = \begin{pmatrix} \overline{f(\zeta)} & 1 \\ f(\zeta) & -1 \end{pmatrix} \begin{pmatrix} 1 & G(K) \\ 0 & F(K) \end{pmatrix}, \tag{2.58}$$

where $f(\zeta) = f(\varrho = 0, \zeta)$ is the axis value of the Ernst potential and $F(K), G(K)$ are integration 'constants' depending on K alone. The particular form of (2.57) is due to the initial condition $\psi = \chi = 1$ for some $\varrho_0 = 0, \zeta = \zeta_0 \in \mathcal{A}^+, \lambda = -1$ (K in the lower sheet), which fixes the second column of Φ in (2.57), cf. (2.50). The first column corresponds to the upper sheet ($\lambda = 1$) and represents a general integral with the two integration 'constants' $F(K), G(K)$ which cannot be specified here. Along $\mathcal{C}, \Phi = \Phi(K)$ does not depend on ϱ and ζ, since A and B vanish because of (2.40) and (2.46). The 'exchange of sheets'[6] along $\mathcal{C}$, together with (2.52), leads to the particular form of Φ on $\mathcal{A}^-$, see Meinel and Neugebauer (1995). The representations (2.57), (2.58) describe the behaviour of ψ and χ in both sheets. Nevertheless, one may wish to consider the whole matrix Φ as a unique function of λ, which is therefore defined on both sheets of the K-surface. From this point of view, Equations (2.57), (2.58) describe Φ on one sheet only (say, on the upper sheet). Its values on the other (lower) sheet follow from (2.52).

---

[6] The path $\mathcal{C}$ can be parameterized by $\varrho = R\sin\vartheta, \zeta = R\cos\vartheta, 0 \le \vartheta \le \pi$ ($R \to \infty$). According to (2.42) this gives $\lambda = \pm\exp(i\vartheta)$.

Combining (2.57) and (2.58) with (2.56), we obtain the axis values in the corotating system,

$$\mathcal{A}^+: \quad \Phi' = \left[1 + i(K-\zeta)\Omega e^{-2U} \begin{pmatrix} -1 & -1 \\ 1 & 1 \end{pmatrix} \right]$$
$$\times \left[\begin{pmatrix} \overline{f(\zeta)} & 1 \\ f(\zeta) & -1 \end{pmatrix} \begin{pmatrix} F(K) & 0 \\ G(K) & 1 \end{pmatrix} \right], \qquad (2.59)$$

$$\mathcal{A}^-: \quad \Phi' = \left[1 + i(K-\zeta)\Omega e^{-2U} \begin{pmatrix} -1 & -1 \\ 1 & 1 \end{pmatrix} \right]$$
$$\times \left[\begin{pmatrix} \overline{f(\zeta)} & 1 \\ f(\zeta) & -1 \end{pmatrix} \begin{pmatrix} 1 & G(K) \\ 0 & F(K) \end{pmatrix} \right], \qquad (2.60)$$

where **1** is the 2×2 unit matrix. At the branch points $K_\text{B} = \zeta$ of K-surfaces belonging to axis values $\varrho = 0, \zeta \in \mathcal{A}^\pm$, ψ and χ must be unique, i.e.

$$\mathcal{A}^+(K_\text{B} = \zeta): \quad \Phi = \begin{pmatrix} \psi & \psi \\ \chi & -\chi \end{pmatrix} = \begin{pmatrix} \overline{f(\zeta)} & 1 \\ f(\zeta) & -1 \end{pmatrix} \begin{pmatrix} F(\zeta) & 0 \\ G(\zeta) & 1 \end{pmatrix}, \qquad (2.61)$$

$$\mathcal{A}^-(K_\text{B} = \zeta): \quad \Phi = \begin{pmatrix} \psi & \psi \\ \chi & -\chi \end{pmatrix} = \begin{pmatrix} \overline{f(\zeta)} & 1 \\ f(\zeta) & -1 \end{pmatrix} \begin{pmatrix} 1 & G(\zeta) \\ 0 & F(\zeta) \end{pmatrix}, \qquad (2.62)$$

whence

$$\mathcal{A}^+: \quad F(\zeta) = \frac{2}{f(\zeta) + \overline{f(\zeta)}}, \quad G(\zeta) = \frac{f(\zeta) - \overline{f(\zeta)}}{f(\zeta) + \overline{f(\zeta)}}, \qquad (2.63)$$

$$\mathcal{A}^-: \quad F(\zeta) = \frac{2f(\zeta)\overline{f(\zeta)}}{f(\zeta) + \overline{f(\zeta)}}, \quad G(\zeta) = \frac{\overline{f(\zeta)} - f(\zeta)}{f(\zeta) + \overline{f(\zeta)}}. \qquad (2.64)$$

Thus, $F(K)$ and $G(K)$ consist in a unique way of analytic continuations of the real and imaginary parts of the axis values of the Ernst potential $f(\zeta)$. Vice versa, $f(\zeta)$ follows from $F(K), G(K)$ for $K = \zeta$. Interestingly, the determinants of Φ and Φ' can be expressed in terms of $\Re f, \Re f'$ and $F(K)$. From (2.41) [$\text{Tr}\Phi_{,z}\Phi^{-1} = (\ln \det \Phi)_{,z}$], (2.46) and (2.57)–(2.60), we have

$$\det \Phi = -2e^{2U} F(K), \qquad \det \Phi' = -2e^{2V} F(K), \qquad (2.65)$$

where $e^{2U} = \Re f$ and $e^{2V} = \Re f'$ [$U = U(\varrho, \zeta), V = V(\varrho, \zeta)$].

We can now interpret the result of (2.59)–(2.64) of the integration of the LP along $\mathcal{A}^+ \mathcal{C} \mathcal{A}^-$: On the regular parts $\mathcal{A}^\pm$ of the symmetry axis, Φ and Φ' can explicitly be expressed in terms of the axis values $f(\zeta)$ of the Ernst potential and its analytic continuations $F(K), G(K)$. To calculate $f(\zeta)$, however, one needs boundary values on $\mathcal{B}$.

2.3 The rigidly rotating disc of dust

Surface of the disc

On the disc, the LP in the corotating system of reference takes the form

$$\mathcal{B}: \quad \mathbf{\Phi}'_{,\varrho} = -\frac{\varrho}{\sqrt{K^2 + \varrho^2(f' + \overline{f'})}} \begin{pmatrix} 0 & \overline{f'}_{,\zeta} \\ f'_{,\zeta} & 0 \end{pmatrix} \mathbf{\Phi}', \quad (2.66)$$

where $\mathbf{\Phi}'$ and f' are the 'corotating' $\mathbf{\Phi}$-matrix and the 'corotating' Ernst potential on the disc. This relation must be considered alongside the 'boundary conditions'

$$\mathbf{\Phi}'(\varrho = 0, \zeta = 0^+, \lambda)|_\mathcal{B} = \mathbf{\Phi}'(\varrho = 0, \zeta = 0^+, \lambda)|_{\mathcal{A}^+},$$

cf. (2.59) and

$$\mathbf{\Phi}'(\varrho = 0, \zeta = 0^-, \lambda)|_\mathcal{B} = \mathbf{\Phi}'(\varrho = 0, \zeta = 0^-, \lambda)|_{\mathcal{A}^-},$$

cf. (2.60). The analysis of this problem will allow for the construction of $F(K)$ and $G(K)$ and, ultimately, via (2.63) and (2.64), for the construction of the axis values $f(\zeta)$ of the Ernst potential.

We first take advantage of the symmetry of the problem, which implies $\overline{f(\varrho, \zeta)} = f(\varrho, -\zeta)$ and connects the ζ-derivatives of f' above ($\zeta = 0^+$) and below ($\zeta = 0^-$) the disc

$$\mathcal{B}: \quad \overline{f'}_{,\zeta}\big|_{\zeta=0^+} = -f'_{,\zeta}\big|_{\zeta=0^-}. \quad (2.67)$$

As a consequence, the LP (2.66) connects the matrix $\overset{A}{\mathbf{\Phi}'}$ above the disc, $\overset{A}{\mathbf{\Phi}'} = \mathbf{\Phi}'(\varrho, \zeta = 0^+, K)$, with the matrix $\overset{B}{\mathbf{\Phi}'}$ below the disc, $\overset{B}{\mathbf{\Phi}'} = \mathbf{\Phi}'(\varrho, \zeta = 0^-, K)$,

$$\mathcal{B}: \quad \overset{A}{\mathbf{\Phi}'} = \begin{pmatrix} 0 & 1 \\ -1 & 0 \end{pmatrix} \overset{B}{\mathbf{\Phi}'} \mathbf{H}(K), \quad (2.68)$$

where the matrix $\mathbf{H}(K)$ (the 'integration constant') does not depend on $\varrho \in \mathcal{B}$. At the rim of the disc we have

$$\overset{A}{\mathbf{\Phi}'}(\varrho_0, 0, K) = \overset{B}{\mathbf{\Phi}'}(\varrho_0, 0, K) = \overset{r}{\mathbf{\Phi}'}. \quad (2.69)$$

Because of (2.68), the rim matrix $\overset{r}{\mathbf{\Phi}'^{-1}} \begin{pmatrix} 0 & -1 \\ 1 & 0 \end{pmatrix} \overset{r}{\mathbf{\Phi}'}$ can be expressed in terms of $\overset{A}{\mathbf{\Phi}'} = \mathbf{\Phi}'(\varrho, \zeta = 0^+, K)$, $\overset{B}{\mathbf{\Phi}'} = \mathbf{\Phi}'(\varrho, \zeta = 0^-, K)$. Note that $\mathbf{\Phi}$ is considered to be a holomorphic function of λ and therefore a function living on the two-sheeted Riemann K-surface of Fig. 2.5. Hence we have to discuss the rim matrix as a function of K on both sheets.

Any Φ multiplied from the right by a matrix function of K is again a solution of the LP. The discussion of the rim matrix simplifies upon using the following renormalization

$$\mathcal{R} = \begin{pmatrix} 0 & 1 \\ -1 & 0 \end{pmatrix} \begin{pmatrix} F & 0 \\ G & 1 \end{pmatrix} \overset{\mathrm{r}}{\Phi}'^{-1} \begin{pmatrix} 0 & -1 \\ 1 & 0 \end{pmatrix} \overset{\mathrm{r}}{\Phi}' \begin{pmatrix} F & 0 \\ G & 1 \end{pmatrix}^{-1} \begin{pmatrix} 0 & 1 \\ -1 & 0 \end{pmatrix}^{-1}. \qquad (2.70)$$

Using (2.59) and (2.60) we obtain

$$\mathcal{R} = \begin{cases} e^{2V_0} \mathcal{M} \mathcal{S}^{-1} & \text{on the upper sheet} \\ -e^{2V_0} \mathcal{S}^{-1} \mathcal{M} & \text{on the lower sheet,} \end{cases} \qquad (2.71)$$

where

$$\mathcal{M} = \begin{pmatrix} G(K) & [G(K)^2 - 1]/F(K) \\ -F(K) & -G(K) \end{pmatrix},$$

$$\mathcal{S} = \begin{pmatrix} f_0 \bar{f_0} - 4\Omega^2 K^2 & i b_0 + 2i\Omega K \\ i b_0 - 2i\Omega K & -1 \end{pmatrix} \qquad (2.72)$$

and

$$f_0 \equiv e^{2V_0} + i b_0 \equiv f(\varrho = 0, \zeta = 0^+). \qquad (2.73)$$

Obviously, $\operatorname{Tr} \mathcal{R} = \operatorname{Tr} \mathcal{R}^{-1} = 0$ and $\mathcal{M}^2 = \mathbf{1}$, whence

$$\operatorname{Tr} \mathcal{M} \mathcal{S}^{-1} = \operatorname{Tr} \mathcal{S} \mathcal{M} = 0. \qquad (2.74)$$

This relation implies one between $F(K)$ and $G(K)$ and, because of (2.63) and (2.64), thus also between the real and imaginary parts of the axis values $f(\zeta)$ of the Ernst potential (Neugebauer and Meinel 1993, 1994).

We next wish to determine $F(K)$ and $G(K)$, which in turn determine $f(\zeta)$. To this end, we consider $\Phi(\varrho, \zeta, \lambda)$ for fixed coordinates ϱ, ζ as a function of λ. We have already used the initial conditions $\psi = \chi = 1$ for some $\varrho = \varrho_1 = 0, \zeta = \zeta_1 \in \mathcal{A}^+$ prescribed in one sheet ($\lambda = -1$) of the K-plane. In principle, the behaviour of Φ in the other sheet and at all points in the ϱ, ζ-plane can be calculated by integrating the LP along a suitable path. However, the coefficients $A(\varrho, \zeta), B(\varrho, \zeta)$ in the LP (2.41) are not explicitly known. Nevertheless, their regular behaviour outside the disc together with the boundary values on the disc provides us with defining properties for Φ. One of them may be taken from Fig. 2.5: Since the domain of the disc, $0 \le \varrho \le \varrho_0, \zeta = 0^\pm$ is a non-vacuum domain, where the LP fails, the matrix Φ cannot remain continuous through the contour(s) $\Gamma: -\varrho_0 \le \Im K \le \varrho_0$ at the branch point pairs $K = i\varrho + 0^\pm, -i\varrho + 0^\pm$, i.e. Φ has a well-defined jump between opposite points along the contour Γ. For fixed coordinate values ϱ, ζ outside the disc ($\varrho, \zeta \notin \mathcal{B}$), a careful discussion would show that $\Phi(\varrho, \zeta, \lambda)$ is a regular function in

2.3 The rigidly rotating disc of dust

λ outside Γ and jumps along Γ, i.e. Φ satisfies a *Riemann–Hilbert problem*. Here we assume Φ to have this behaviour and are justified in our assumption in that we obtain the solution to the problem.

Let us now consider the jump $\Phi_+^{-1}\Phi_-$, where the signs refer to the two sides of Γ, cf. Fig. 2.5. The LP tells us that $\Phi_+^{-1}\Phi_-$ does not depend on the coordinates and is therefore a function $\mathcal{D}$ of the contour alone,

$$\Phi_+^{-1}\Phi_- = \mathcal{D}_u(K), \qquad K \in \Gamma_u,$$
$$\Phi_+^{-1}\Phi_- = \mathcal{D}_l(K), \qquad K \in \Gamma_l, \tag{2.75}$$

where u denotes the upper and l the lower sheet. Since the jump contours Γ_u, Γ_l and the jump matrices $\mathcal{D}_u, \mathcal{D}_l$ are the same for all values of ϱ, ζ (i.e. for all Riemann surfaces with different branch points), we can express $\mathcal{D}_u$ and $\mathcal{D}_l$ in terms of the axis values of Φ,

$$\mathcal{D}_u(K) = \begin{pmatrix} F_+ & 0 \\ G_+ & 1 \end{pmatrix}^{-1} \begin{pmatrix} F_- & 0 \\ G_- & 1 \end{pmatrix}, \qquad K \in \Gamma_u. \tag{2.76}$$

A similar relation for $\mathcal{D}_l$ may be obtained via (2.52). As a consequence of (2.75), (2.76) the matrix $\Phi \begin{pmatrix} F & 0 \\ G & 1 \end{pmatrix}^{-1}$ does not jump along Γ_u. Because of (2.56), the same holds for $\Phi' \begin{pmatrix} F & 0 \\ G & 1 \end{pmatrix}^{-1}$ and, finally, for $\mathcal{R}$ as defined in (2.70). Consider now the Riemann K-surface corresponding to the disc's rim $\varrho = \varrho_0, \zeta = 0$. The cut between the branch points $K_B = \pm i\varrho_0$ coincides with the contour Γ_u, Γ_l which are on the two 'bridges' connecting crosswise the upper with the lower sheet. Since $\mathcal{R}$ does not jump on Γ_u, we have, according to (2.71), $(\mathcal{MS}^{-1})_- = -(\mathcal{S}^{-1}\mathcal{M})_+$. Though $\mathcal{R}$ does not jump, F and G do jump, cf. (2.76). Note that $F(K)$ and $G(K)$ are unique functions of K. Hence, there is only one contour $\Gamma : \Re K = 0, -\varrho_0 \leq \Im K \leq \varrho_0$ where $\mathcal{M}$ does jump. Since Φ is analytic outside Γ_u, Γ_l, the matrix $\mathcal{M}$ must be analytic outside Γ. Thus we obtain $F(K)$ and $G(K)$ from the Riemann–Hilbert problem

$$K \in \Gamma: \quad \mathcal{SM}_- = -\mathcal{M}_+\mathcal{S},$$
$$K \notin \Gamma: \quad \mathcal{M}(K) \text{ analytic in } K, \tag{2.77}$$

$\mathcal{S}$ and $\mathcal{M}$ as in (2.72). (Note that the elements of $\mathcal{S}$, which are polynomials, and the elements of $\mathcal{S}^{-1}$, which are rational functions in K, do not jump along Γ.) There is no jump at the end points of the contour $K = \pm i\varrho_0, \mathcal{M}(\pm i\varrho_0)_- = \mathcal{M}(\pm i\varrho_0)_+$. As a consequence, one obtains $\text{Tr }\mathcal{S}(\pm i\varrho_0) = 0$, i.e. the parameter relation

$$f_0\bar{f}_0 + 4\Omega^2\varrho_0^2 = 1. \tag{2.78}$$

It turns out that the Riemann–Hilbert problem (2.77) has a unique solution $\mathcal{M}(K)$ in the parameter region[7]

$$0 < \mu \equiv 2\Omega^2 e^{-2V_0}\varrho_0^2 < \mu_0 = 4.62966184\ldots \qquad (2.79)$$

An important step on the way to this solution is the diagonalization of $\mathcal{S}$. Finally, one obtains $F(K), G(K)$ and the axis values of the Ernst potential $f(\zeta)$ in terms of elliptic theta functions. We need not go along this route. As we shall see in the following, we can use the Riemann–Hilbert problem (2.77) to formulate a more general Riemann–Hilbert problem, which will yield the complete disc of dust solution in terms of hyperelliptic theta functions.

Ernst potential everywhere

In order to construct the Φ-matrix for arbitrary values of ϱ, ζ and λ, let us return to the Riemann–Hilbert problem (2.77). As we have seen, the matrix $\Phi \begin{pmatrix} F & 0 \\ G & 1 \end{pmatrix}^{-1}$ does not jump along Γ_{u}. Analogously, $\Phi \begin{pmatrix} 1 & G \\ 0 & F \end{pmatrix}^{-1}$ does not jump along Γ_{l}. The images Γ_λ of Γ_{u} and $\Gamma_{-\lambda}$ of Γ_{l} inherit these properties, which is essential to the following deductions.

To formulate a Riemann–Hilbert problem in the λ-plane, we define two matrices,

$$\begin{aligned}
\mathcal{L} &:= \Phi \begin{pmatrix} 1 & G \\ 0 & F \end{pmatrix}^{-1} \begin{pmatrix} 1 & 0 \\ 0 & -1 \end{pmatrix} \mathcal{M} \begin{pmatrix} 1 & 0 \\ 0 & -1 \end{pmatrix} \begin{pmatrix} 1 & G \\ 0 & F \end{pmatrix} \Phi^{-1} \\
&= \Phi \begin{pmatrix} F & 0 \\ G & 1 \end{pmatrix}^{-1} \begin{pmatrix} 0 & -1 \\ 1 & 0 \end{pmatrix} \mathcal{M} \begin{pmatrix} 0 & 1 \\ -1 & 0 \end{pmatrix} \begin{pmatrix} F & 0 \\ G & 1 \end{pmatrix} \Phi^{-1} \qquad (2.80) \\
&= \Phi \begin{pmatrix} 0 & 1 \\ 1 & 0 \end{pmatrix} \Phi^{-1},
\end{aligned}$$

$$\begin{aligned}
\mathcal{Q} &:= \mathrm{e}^{-2V_0} \Phi \begin{pmatrix} 1 & G \\ 0 & F \end{pmatrix}^{-1} \begin{pmatrix} 1 & 0 \\ 0 & -1 \end{pmatrix} (\mathcal{S} + w\mathbf{1}) \begin{pmatrix} 1 & 0 \\ 0 & -1 \end{pmatrix} \begin{pmatrix} 1 & G \\ 0 & F \end{pmatrix} \Phi^{-1} \\
&= \mathrm{e}^{-2V_0} \Phi \begin{pmatrix} F & 0 \\ G & 1 \end{pmatrix}^{-1} \begin{pmatrix} 0 & -1 \\ 1 & 0 \end{pmatrix} (\mathcal{S} + w\mathbf{1}) \begin{pmatrix} 0 & 1 \\ -1 & 0 \end{pmatrix} \begin{pmatrix} F & 0 \\ G & 1 \end{pmatrix} \Phi^{-1},
\end{aligned} \qquad (2.81)$$

where

$$w = -\frac{1}{2}\mathrm{Tr}\,\mathcal{S} = 2\Omega^2(K^2 + \varrho_0^2). \qquad (2.82)$$

---

[7] In this section the symbol μ denotes a parameter defined in (2.79), and should not be confused with the mass-density. The meaning of μ_0 will be discussed in Subsection 2.3.3.

2.3 The rigidly rotating disc of dust

Here we have made use of the parameter relation (2.78). Since $\mathcal{S}$ and w are polynomials in K, and therefore rational functions in λ, the matrix $\mathcal{Q}$ does not jump at all. Taking the asymptotics of $\mathcal{S}$ and w into account, $\mathcal{Q}$ must have the following structure in λ:

$$\mathcal{Q} = (K + \mathrm{i}z)^2 \begin{pmatrix} q_1 & q_2 \\ q_3 & -q_2 \end{pmatrix}, \tag{2.83}$$

with the polynomials

$$q_1 = k\lambda + l\lambda^3, \quad q_2 = m + n\lambda^2 + p\lambda^4, \quad q_3 = q + r\lambda^2 + s\lambda^4, \tag{2.84}$$

where k, l, m, n, p, q, r and s are functions of ϱ, ζ alone. From the definitions (2.80), (2.81) and condition (2.74), we can derive

$$\mathcal{Q}\mathcal{L} = -\mathcal{L}\mathcal{Q}, \tag{2.85}$$

whereas the particular Riemann–Hilbert problem (2.77) has the continuation

$$\lambda \in \Gamma_\lambda : \qquad (\mathcal{Q} + \mathrm{e}^{-2V_0}w\mathbf{1})\mathcal{L}_- = -\mathcal{L}_+(\mathcal{Q} + \mathrm{e}^{-2V_0}w\mathbf{1}),$$

$$\lambda \in \Gamma_{-\lambda} : \qquad (\mathcal{Q} - \mathrm{e}^{-2V_0}w\mathbf{1})\mathcal{L}_- = -\mathcal{L}_+(\mathcal{Q} - \mathrm{e}^{-2V_0}w\mathbf{1}), \tag{2.86}$$

$$\lambda \notin \Gamma_\lambda, \Gamma_{-\lambda} : \qquad \mathcal{L} \text{ analytic in } \lambda.$$

The following solution of the regular Riemann–Hilbert problem (2.86) is based on the diagonalization of $\mathcal{Q}$.

We consider a function Ψ defined by

$$\Psi := \frac{1}{\sqrt{w^2 + \mathrm{e}^{4V_0}}} \ln \frac{\hat{\mathcal{L}}_{22} + \sqrt{1 + w^2 \mathrm{e}^{-4V_0}}\hat{\mathcal{L}}_{21}}{\hat{\mathcal{L}}_{22} - \sqrt{1 + w^2 \mathrm{e}^{-4V_0}}\hat{\mathcal{L}}_{21}}, \tag{2.87}$$

where

$$\hat{\mathcal{L}} = \mathcal{L} \begin{pmatrix} 1 & \mathcal{Q}_{11} \\ 0 & \mathcal{Q}_{21} \end{pmatrix}. \tag{2.88}$$

Note that Ψ has no branch points at the zeros $K_1, K_2, \overline{K}_1 = -K_2$ and $\overline{K}_2 = -K_1$ of $w^2 + \mathrm{e}^{4V_0}$,

$$K_1^2 = \varrho_0^2 \frac{\mathrm{i} - \mu}{\mu}, \quad K_2^2 = \varrho_0^2 \frac{\mathrm{i} + \mu}{\mu} \quad [\Re K_1 < 0, \quad \Re K_2 > 0, \quad \mu \text{ as in (2.79)}], \tag{2.89}$$

since Ψ is unaffected by a change in the sign of $\sqrt{w^2 + \mathrm{e}^{4V_0}}$. It is an odd function of λ, thus vanishing at $\lambda = 0$ and at $\lambda = \infty$. Therefore, the function

$$\hat{\Psi} := \Psi/[\lambda(K + \mathrm{i}z)] = \Psi/\sqrt{(K - \mathrm{i}\bar{z})(K + \mathrm{i}z)} \tag{2.90}$$

is a unique function of K, which is characterized by the following properties:

(i) There is a jump along Γ, which because of (2.86) reads

$$\hat{\Psi}_- = \hat{\Psi}_+ + \frac{2}{\sqrt{(K-i\bar{z})(K+iz)}\sqrt{w^2+e^{4V_0}}} \ln \frac{\sqrt{w^2+e^{4V_0}}+w}{\sqrt{w^2+e^{4V_0}}-w}. \quad (2.91)$$

(ii) Because of

$$\hat{\mathcal{L}}_{21}^2(1+w^2 e^{-4V_0}) - \hat{\mathcal{L}}_{22}^2 = \mathcal{Q}_{21}^2, \quad (2.92)$$

$$\mathcal{Q}_{21} = -\frac{2f\Omega^2 e^{-2V_0}}{f+\bar{f}}(K-K_a)(K-K_b), \quad (2.93)$$

the behaviour for $K \to K_{a/b}$ is given by

$$\hat{\Psi} \to \frac{\pm 2}{\sqrt{(K_{a/b}-i\bar{z})(K_{a/b}+iz)(w_{a/b}^2+e^{4V_0})}} \ln(K-K_{a/b}) \quad \text{as} \quad K \to K_{a/b}. \quad (2.94)$$

(The ambiguity of sign can be compensated for by the square root.)

(iii) Because of the definitions of $\mathcal{Q}$ and $\mathcal{L}$, the behaviour for $K \to \infty$ is given by

$$\hat{\Psi} \to \frac{\ln f}{\Omega^2 K^3} \quad \text{as} \quad K \to \infty. \quad (2.95)$$

A representation of $\hat{\Psi}$ that possesses all these properties is

$$\hat{\Psi} = \frac{1}{\pi i} \int_{-i\varrho_0}^{i\varrho_0} \frac{\ln[\sqrt{w'^2+e^{4V_0}}+w']/[\sqrt{w'^2+e^{4V_0}}-w']}{\sqrt{(K'-i\bar{z})(K'+iz)}\sqrt{w'^2+e^{4V_0}}(K'-K)} dK'$$

$$-2\int_{K_1}^{K_a} \frac{1}{\sqrt{(K'-i\bar{z})(K'+iz)(w'^2+e^{4V_0})}(K'-K)} dK' \quad (2.96)$$

$$-2\int_{K_2}^{K_b} \frac{1}{\sqrt{(K'-i\bar{z})(K'+iz)(w'^2+e^{4V_0})}(K'-K)} dK',$$

where K_a and K_b have to be determined such that $\hat{\Psi} = \mathcal{O}(K^{-3})$. The lower limits of integration in the last two integrals have been fixed to obtain the correct result in the Newtonian limit $\mu \to 0$ where $K_a/K_1 = 1 + \mathcal{O}(\mu^2)$ and $K_b/K_2 = 1 + \mathcal{O}(\mu^2)$. For a systematic post-Newtonian expansion of the solution, see Petroff and Meinel (2001). Note that the last two terms in Equation (2.96) may also be interpreted as follows,

$$2\left(\int_{K_1}^{K_a} + \int_{K_2}^{K_b}\right) = 2\left(\int_{K_a}^{K_1}\{-\} + \int_{K_2}^{K_b}\right) = \int_{K_a}^{K_b}\{1\} + \int_{K_a}^{K_b}\{2\}, \qquad (2.97)$$

showing that nothing special happens at K_1 and K_2. In this symbolic notation, $\{-\}$ indicates that the square root is meant to have the opposite sign from that of the first term; $\{1\}$ and $\{2\}$ denote different paths in the complex K-plane, which are chosen such that the closed integral

$$\oint = \int_{K_a}^{K_b}\{1\} - \int_{K_a}^{K_b}\{2\} = 2\int_{K_1}^{K_2} \qquad (2.98)$$

is performed around a contour enclosing the branch points K_1 and K_2 of $\sqrt{w^2 + e^{4V_0}}$. In the subsequent formulae we normalize K by introducing

$$X = \frac{K}{\varrho_0}, \quad X_{a/b} = \frac{K_{a/b}}{\varrho_0}, \quad X_{1/2} = \frac{K_{1/2}}{\varrho_0}. \qquad (2.99)$$

According to (2.95), an asymptotic expansion of Equation (2.96) for $X \to \infty$ ($K \to \infty$) leads to

$$\ln f = \mu\left[\int_{X_1}^{X_a} \frac{X^2 dX}{W} + \int_{X_2}^{X_b} \frac{X^2 dX}{W} - \int_{-i}^{i} \frac{hX^2 dX}{W_1}\right], \qquad (2.100)$$

$$\int_{X_1}^{X_a} \frac{dX}{W} + \int_{X_2}^{X_b} \frac{dX}{W} = \int_{-i}^{i} \frac{h dX}{W_1}, \quad \int_{X_1}^{X_a} \frac{X dX}{W} + \int_{X_2}^{X_b} \frac{X dX}{W} = \int_{-i}^{i} \frac{hX dX}{W_1}, \qquad (2.101)$$

where the lower limits of integration X_1, X_2 are given by

$$X_1^2 = \frac{i - \mu}{\mu}, \quad X_2^2 = -\frac{i + \mu}{\mu} \quad (\Re X_1 < 0, \quad \Re X_2 > 0), \qquad (2.102)$$

whereas the upper limits X_a, X_b must be calculated from the integral equations (2.101). Here we have introduced the abbreviations

$$W = W_1 W_2, \quad W_1 = \sqrt{(X - \zeta/\varrho_0)^2 + (\varrho/\varrho_0)^2},$$

$$W_2 = \sqrt{1 + \mu^2(1 + X^2)^2} \qquad (2.103)$$

and
$$h = \frac{\ln\left(\sqrt{1+\mu^2(1+X^2)^2} + \mu(1+X^2)\right)}{\pi i \sqrt{1+\mu^2(1+X^2)^2}}. \tag{2.104}$$

The third integral in (2.100) as well as the integrals on the right hand sides in (2.101) have to be taken along the imaginary axis in the complex X-plane with h and W_1 fixed according to $\Re W_1 < 0$ (for ϱ, ζ outside the disc) and $\Re h = 0$. The task of calculating the upper limits X_a, X_b in (2.101) from

$$u = \int_{-i}^{i} \frac{h\,dX}{W_1}, \quad v = \int_{-i}^{i} \frac{hX\,dX}{W_1} \tag{2.105}$$

is known as Jacobi's inversion problem. Göpel (1847) and Rosenhain (1850) were able to express the hyperelliptic functions $X_a(u,v)$ and $X_b(u,v)$ in terms of (hyperelliptic) theta functions. Later on it turned out that even the first two integrals in (2.100) can be expressed by theta functions in u and v! A detailed introduction to the related mathematical theory which was founded by Riemann and Weierstrass may be found in Stahl (1896), Krazer (1903) and Belokolos et al. (1994). The representation of the Ernst potential (2.100) in terms of theta functions can be taken from Stahl's book, see Stahl (1896), p. 311, Equation (5). Here is the result: Defining a theta function $\vartheta(x,y;p,q,\alpha)$ by

$$\vartheta(x,y;p,q,\alpha) := \sum_{m=-\infty}^{\infty}\sum_{n=-\infty}^{\infty} (-1)^{m+n} p^{m^2} q^{n^2} e^{2mx+2ny+4mn\alpha}, \tag{2.106}$$

one can reformulate the expressions (2.100), (2.101) to give

$$f = \frac{\vartheta(\alpha_0 u + \alpha_1 v - C_1, \beta_0 u + \beta_1 v - C_2; p,q,\alpha)}{\vartheta(\alpha_0 u + \alpha_1 v + C_1, \beta_0 u + \beta_1 v + C_2; p,q,\alpha)} e^{-(\gamma_0 u + \gamma_1 v + \mu w)}, \tag{2.107}$$

with u and v as in (2.105) and

$$w = \int_{-i}^{i} \frac{hX^2\,dX}{W_1}. \tag{2.108}$$

The normalization parameters $\alpha_0, \alpha_1; \beta_0, \beta_1; \gamma_0, \gamma_1$, the moduli p, q, α of the theta function and the quantities C_1, C_2 are defined on the two sheets of the hyperelliptic Riemann surface related to

$$W = \mu\sqrt{(X-X_1)(X-\bar{X}_1)(X-X_2)(X-\bar{X}_2)(X-i\bar{z}/\varrho_0)(X+iz/\varrho_0)}, \tag{2.109}$$

2.3 The rigidly rotating disc of dust

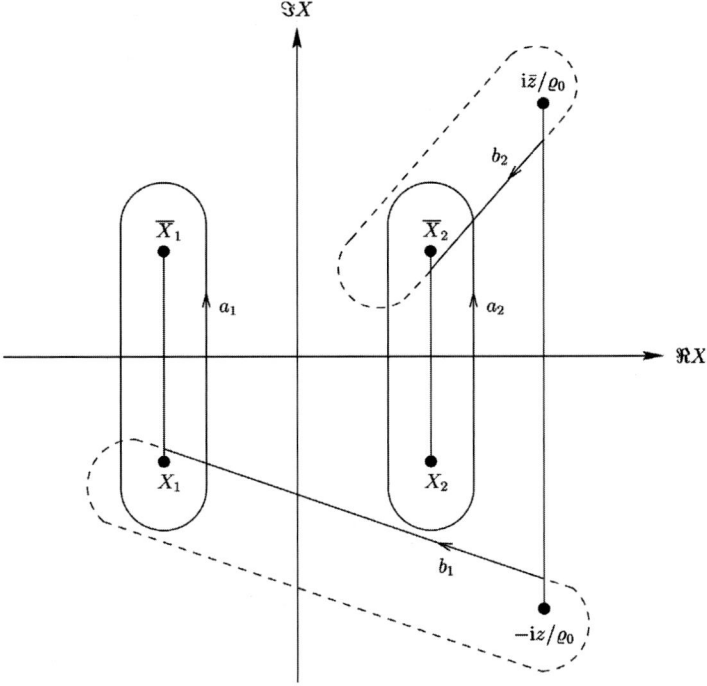

Fig. 2.6. Riemann surface with cuts between the branch points X_1 and $\bar{X}_1$, X_2 and $\bar{X}_2$, $-\mathrm{i}z/\varrho_0$ and $\mathrm{i}\bar{z}/\varrho_0$. Also shown are the four periods a_i and b_i ($i=1,2$). Solid/dashed lines belong to the upper/lower sheet defined by $W \to \pm \mu X^3$ as $X \to \infty$.

see Fig. 2.6. There are two normalized Abelian differentials of the first kind

$$\mathrm{d}\omega_1 = \alpha_0 \frac{\mathrm{d}X}{W} + \alpha_1 \frac{X\,\mathrm{d}X}{W}, \tag{2.110}$$

$$\mathrm{d}\omega_2 = \beta_0 \frac{\mathrm{d}X}{W} + \beta_1 \frac{X\,\mathrm{d}X}{W}, \tag{2.111}$$

defined by

$$\oint_{a_m} \mathrm{d}\omega_n = \pi\mathrm{i}\,\delta_{mn} \quad (m=1,2;\,n=1,2). \tag{2.112}$$

Equation (2.112) consists of four linear, algebraic equations and yields the four parameters $\alpha_0, \alpha_1, \beta_0, \beta_1$ in terms of integrals extending over the closed (deformable) curves a_1, a_2. It can be shown that there is one normalized Abelian differential of the third kind,

$$\mathrm{d}\omega = \gamma_0 \frac{\mathrm{d}X}{W} + \gamma_1 \frac{X\,\mathrm{d}X}{W} + \mu \frac{X^2\,\mathrm{d}X}{W}, \tag{2.113}$$

with vanishing a-periods

$$\oint_{a_j} d\omega = 0 \quad (j = 1, 2). \tag{2.114}$$

This equation defines γ_0 and γ_1, again via a linear, algebraic system. The Riemann matrix

$$(B_{ij}) = \begin{pmatrix} \ln p & 2\alpha \\ 2\alpha & \ln q \end{pmatrix} \quad (i = 1, 2; j = 1, 2) \tag{2.115}$$

(with negative definite real part) is given by

$$B_{ij} = \oint_{b_i} d\omega_j \tag{2.116}$$

and defines the moduli p, q, α of the theta function (2.106). Finally, the quantities C_1, C_2 can be calculated by

$$C_i = -\int_{-iz/\varrho_0}^{\infty^+} d\omega_i \quad (i = 1, 2), \tag{2.117}$$

where '+' denotes the upper sheet. Obviously, all the quantities entering the theta functions and the exponential function in (2.107) can be expressed in terms of well-defined integrals and depend on the three parameters $\varrho/\varrho_0, \zeta/\varrho_0, \mu$. The corresponding 'tables' for $\alpha_i, \beta_i, \gamma_i, C_i, B_{ij}, u, v$ and w can easily be calculated by numerical integration. Fortunately, theta series like (2.106) converge rapidly. For $0 < \mu < \mu_0$, the solution (2.107) is analytic *everywhere* outside the disc, even at the rings $-iz/\varrho_0 = X_1, X_2$. The limit $\mu \to \mu_0$ will be discussed in detail in Subsection 2.3.5. Note that for $\mu > \mu_0$, the Ernst potential (2.107) is no longer singularity-free outside the disc. This corresponds to the fact that the boundary value problem (2.39) has a unique solution, since the range $0 < \mu < \mu_0$ already covers the full range $0 > V_0 > -\infty$, as will be shown in Subsection 2.3.3.

The complete metric

The metric functions e^{2U}, a and e^{2k} calculated from the Ernst potential (2.107) according to (1.44) and (1.110) are given as follows:

$$e^{2U} = \frac{\vartheta(\mathbf{c})\vartheta^*(\mathbf{c})\vartheta(\mathbf{a})\vartheta^*(\mathbf{a})}{\vartheta(\mathbf{0})\vartheta^*(\mathbf{0})\vartheta(\mathbf{a}+\mathbf{c})\vartheta^*(\mathbf{a}+\mathbf{c})} e^{-(\gamma_0 u + \gamma_1 v + \mu w)}, \qquad (2.118)$$

$$1 + \frac{(1+\Omega a)e^{2U}}{\Omega\varrho} = \frac{\vartheta(\mathbf{0})\vartheta^*(\mathbf{0})\vartheta(\mathbf{a}+2\mathbf{c})\vartheta^*(\mathbf{a})}{\vartheta(\mathbf{c})\vartheta^*(\mathbf{0})\vartheta(\mathbf{a}+\mathbf{c})\vartheta^*(\mathbf{a}+\mathbf{c})}, \qquad (2.119)$$

$$e^{2k(\varrho,\zeta)} = \frac{\kappa(\varrho,\zeta)}{\kappa(0,0)} \qquad (2.120)$$

with

$$\kappa(\varrho,\zeta) = \frac{\vartheta(\mathbf{a})\vartheta^*(\mathbf{a})}{\vartheta(\mathbf{0})\vartheta^*(\mathbf{0})} \exp\left[2k_0 - \frac{1}{2}\sum_{i,k=1}^{2} a_i a_k \frac{\partial^2 \ln \vartheta(\mathbf{x})\vartheta^*(\mathbf{x})}{\partial x_i \partial x_k}\bigg|_{\mathbf{x}=\mathbf{0}}\right], \qquad (2.121)$$

where

$$2k_0 = \frac{\mu^2}{4} \int_{-i}^{i}\int_{-i}^{i} (X-X_1)(X-X_2)(X'+X_1)(X'+X_2)$$

$$\times \frac{(\lambda-\lambda')^2}{\lambda\lambda'} \frac{h(X)h(X')}{(X-X')^2} dX \, dX', \qquad (2.122)$$

$$\lambda = \sqrt{\frac{X-i\bar{z}/\varrho_0}{X+iz/\varrho_0}}, \qquad \lambda' = \sqrt{\frac{X'-i\bar{z}/\varrho_0}{X'+iz/\varrho_0}}, \qquad (2.123)$$

$$\vartheta(\mathbf{x}) = \vartheta(\mathbf{x};p,q,\alpha) = \vartheta(x_1,x_2;p,q,\alpha), \qquad (2.124)$$

$$\vartheta^*(\mathbf{x}) = \vartheta\left(x_1+\frac{i\pi}{2},x_2+\frac{i\pi}{2};p,q,\alpha\right), \qquad (2.125)$$

$$\mathbf{a} = (a_1,a_2) = (\alpha_0 u + \alpha_1 v, \beta_0 u + \beta_1 v), \quad \mathbf{0} = (0,0), \quad \mathbf{c} = (C_1, C_2). \qquad (2.126)$$

2.3.3 Mathematical discussion of the solution

The aim of this subsection is to perform a careful analysis of the mathematical properties of the solution of the rigidly rotating disc of dust in terms of theta functions. The main focus of our attention will be the Ernst potential (2.107), but we shall also consider Equations (2.119) and (2.120) for the complete metric. We begin by presenting a general method permitting the use of these equations in a direct numerical calculation (Kleinwächter 2001). The result is theta formulae that contain only definite real integrals as arguments. Besides providing for easy numerical applicability, all the related equations are well suited for specializations to regions of special interest, such as the axis and the disc itself as well as the whole symmetry plane. An alternative representation of the solution can be found in Appendix 4.

Throughout this and the next subsection, we shall restrict ourselves to the region $\zeta > 0$. The symmetry relations $U(\varrho, -\zeta) = U(\varrho, \zeta), a(\varrho, -\zeta) = a(\varrho, \zeta)$, $k(\varrho, -\zeta) = k(\varrho, \zeta)$ and $f(\varrho, -\zeta) = \overline{f(\varrho, \zeta)}$, see (1.88) and (1.111), can be used to calculate the solution for $\zeta < 0$. Note that the metric functions U, a and k are continuous at $\zeta = 0$ but the imaginary part of the Ernst potential b is discontinuous at the disc: $b(\varrho, \zeta = 0^-) = -b(\varrho, \zeta = 0^+)$. The same holds for the function v of (2.105).

Discussion of the theta formula for the Ernst potential

In what follows, the elliptic theta functions $\vartheta_1, \ldots, \vartheta_4$ introduced by Jacobi and some of the ultra-elliptic theta functions $\vartheta_{n,k}(n, k = 1, 2, 3, 4)$ introduced by Rosenhain are used. Moreover Jacobi's elliptic functions sn, cn, dn, the elliptic integrals F and E, the Jacobian zeta function Z and Heuman's lambda function Λ_0 are used. All these functions are defined in Appendix 2 and some basic properties are provided there.

In this notation, (2.107) for the Ernst potential reads

$$f = e^{-(\gamma_0 u + \gamma_1 v + \mu w)} \frac{\vartheta_{4,4}(\alpha_0 u + \alpha_1 v - C_1, \beta_0 u + \beta_1 v - C_2; B_{11}, B_{22}, B_{12})}{\vartheta_{4,4}(\alpha_0 u + \alpha_1 v + C_1, \beta_0 u + \beta_1 v + C_2; B_{11}, B_{22}, B_{12})}. \quad (2.127)$$

Before starting, we recapitulate the meaning of the arguments in the above formula making use of the normalized coordinates

$$x := \varrho/\varrho_0, \quad y := \zeta/\varrho_0. \quad (2.128)$$

The quantities $\alpha_0, \alpha_1, \beta_0, \beta_1, \gamma_0, \gamma_1, C_1, C_2, B_{11}, B_{22}$ and B_{12} of the 'theta formula' for f depend via ultra-elliptic line integrals on the above normalized coordinates and on the parameter μ. The functions u, v and w, which depend on x, y and μ, are given by[8]

$$u = \int_{-i}^{i} \frac{h \, dX}{W_1}, \quad v = \int_{-i}^{i} \frac{h X \, dX}{W_1}, \quad w = \int_{-i}^{i} \frac{h X^2 \, dX}{W_1}, \quad (2.129)$$

---

[8] The convention for the root W_1 is $\Re W_1 < 0$ for $(x, y) \notin \{(r, 0) : 0 \leq r \leq 1\}$.

2.3 The rigidly rotating disc of dust

with

$$h = \frac{\ln\left(\sqrt{1+\mu^2(1+X^2)^2} + \mu(1+X^2)\right)}{i\pi\sqrt{1+\mu^2(1+X^2)^2}}, \quad W_1 = \sqrt{(X-y)^2 + x^2}. \quad (2.130)$$

The normalization parameters $\alpha_0, \alpha_1, \beta_0, \beta_1, \gamma_0, \gamma_1$, the moduli B_{mn} of the theta functions and the quantities C_1, C_2 are defined via integrals on the two sheets of the hyperelliptic Riemann surface (see Fig. 2.6) related to

$$W = \mu\sqrt{(X-X_1)(X-\overline{X_1})(X-X_2)(X-\overline{X_2})(X-y-ix)(X-y+ix)} \quad (2.131)$$

with

$$X_1 = -\frac{1}{\sqrt{2\mu}}\left(\sqrt{\sqrt{1+\mu^2}-\mu} + i\sqrt{\sqrt{1+\mu^2}+\mu}\right)$$

and

$$X_2 = -\overline{X_1}.$$

The upper sheet of the surface is characterized by $W \to +\mu X^3$ for $X \to \infty$. The two normalized Abelian differentials of the first kind are defined by

$$d\omega_1 := \alpha_0 \frac{dX}{W} + \alpha_1 \frac{X\,dX}{W}, \quad d\omega_2 := \beta_0 \frac{dX}{W} + \beta_1 \frac{X\,dX}{W} \quad (2.132)$$

with

$$\oint_{a_m} d\omega_n \stackrel{!}{=} i\pi \delta_{mn} \quad (m=1,2; n=1,2). \quad (2.133)$$

The values of γ_0 and γ_1 have to be determined by integrating over an Abelian differential of the third kind

$$d\omega := \gamma_0 \frac{dX}{W} + \gamma_1 \frac{X\,dX}{W} + \mu \frac{X^2\,dX}{W} \quad (2.134)$$

with the requirement:

$$\oint_{a_m} d\omega \stackrel{!}{=} 0 \quad (m=1,2). \quad (2.135)$$

The remaining quantities are defined to be

$$B_{mn} := \oint_{b_m} d\omega_n \quad \text{and} \quad C_m := \int_{\infty^+}^{y-ix} d\omega_m \quad (m=1,2; n=1,2). \quad (2.136)$$

Analytical treatment of limiting cases

Our first goal is to calculate all the arguments once and for all so that for applications it is not necessary to consider the Riemann surface and Abelian differentials any more. Using these results, the theta formula will then be rewritten. With the notation

$$\xi_1 := -\frac{1}{\sqrt{2\mu}}\sqrt{\sqrt{1+\mu^2}-\mu}, \quad \eta_1 := \frac{1}{\sqrt{2\mu}}\sqrt{\sqrt{1+\mu^2}+\mu}, \quad (2.137)$$

$$\xi_2 := \frac{1}{\sqrt{2\mu}}\sqrt{\sqrt{1+\mu^2}-\mu}, \quad \eta_2 := \frac{1}{\sqrt{2\mu}}\sqrt{\sqrt{1+\mu^2}+\mu}, \quad (2.138)$$

$$\xi_3 := y, \quad \eta_3 := x, \quad (2.139)$$

the cuts in the Riemann surface (Fig. 2.6) are denoted by $V(\xi_1, \eta_1)$, $V(\xi_2, \eta_2)$ and $V(\xi_3, \eta_3)$ with

$$V(\xi_i, \eta_i) := \{\xi_i + \mathrm{i}s\,\eta_i : s \in [-1, 1]\}. \quad (2.140)$$

Let us carefully consider the behaviour of the 'fundamental root' (2.131). First we introduce the 'elementary root' for the complex quantity $X = r + \mathrm{i}s$ jumping at the cut $V(a, b)$

$$\mathcal{W}(r, s; a, b) := (X - a)\sqrt{1 + \frac{b^2}{(X-a)^2}} \quad (r, s \text{ and } a, b \text{ real}), \quad (2.141)$$

where the choice $\Re\sqrt{\ldots} \geq 0$ is made. The fundamental root (2.131) can be built up from (2.141) using the definitions (2.137)–(2.139) and (2.140)

$$W(r, s; \mu, x, y) = \mu \mathcal{W}(r, s; \xi_1, \eta_1)\mathcal{W}(r, s; \xi_2, \eta_2)\mathcal{W}(r, s; \xi_3, \eta_3). \quad (2.142)$$

The radicand $\mathcal{R}(r, s; a, b)$ of (2.141) is given by

$$1 + \frac{b^2}{(X-a)^2} = \mathcal{R}_R(r, s; a, b) + \mathrm{i}\,\mathcal{R}_I(r, s; a, b), \quad (2.143)$$

$$\mathcal{R}_R(r, s; a, b) \equiv 1 + \frac{b^2\left[(r-a)^2 - s^2\right]}{\left[(r-a)^2 + s^2\right]^2}, \quad (2.144)$$

$$\mathcal{R}_I(r, s; a, b) \equiv -2s\frac{(r-a)b^2}{\left[(r-a)^2 + s^2\right]^2}. \quad (2.145)$$

2.3 The rigidly rotating disc of dust

With the notation $\mathcal{W} = (X-a)(\mathcal{S}_R + i\mathcal{S}_I)$, one finds[9]

$$\mathcal{S}_R(r,s;a,b) \equiv \frac{1}{\sqrt{2}}\sqrt{\sqrt{\mathcal{R}_R^2 + \mathcal{R}_I^2} + \mathcal{R}_R}, \qquad (2.146)$$

$$\mathcal{S}_I(r,s;a,b) \equiv \text{sign}(\mathcal{R}_I)\frac{1}{\sqrt{2}}\sqrt{\sqrt{\mathcal{R}_R^2 + \mathcal{R}_I^2} - \mathcal{R}_R}, \qquad (2.147)$$

$$\text{sign}(\mathcal{R}_I) = -\text{sign}(s)\,\text{sign}(r-a). \qquad (2.148)$$

The definition

$$g_\pm(r,s;a,b) := \frac{1}{\sqrt{2}}\sqrt{\sqrt{\mathcal{R}_R^2 + \mathcal{R}_I^2} \pm \mathcal{R}_R} \qquad (2.149)$$

$$= \frac{1}{\sqrt{2}}\left\{\frac{\sqrt{\left[(r-a)^2 + s^2 + b^2\right]^2 - 4b^2 s^2}}{(r-a)^2 + s^2} \pm \left\{1 + \frac{b^2\left[(r-a)^2 - s^2\right]}{\left[(r-a)^2 + s^2\right]^2}\right\}\right\}^{\frac{1}{2}}$$

leads to the following representation of the elementary root

$$\mathcal{W}(r,s;a,b) = \mathcal{W}_R(r,s;a,b) + i\mathcal{W}_I(r,s;a,b), \qquad (2.150)$$

$$\mathcal{W}_R(r,s;a,b) = \text{sign}(r-a)\left[|r-a|g_+(r,s;a,b) + |s|g_-(r,s;a,b)\right],$$

$$\mathcal{W}_I(r,s;a,b) = \text{sign}(s)\left[|s|g_+(r,s;a,b) - |r-a|g_-(r,s;a,b)\right]. \qquad (2.151)$$

From these equations, one can work out the following properties for $\mathcal{W}$ and W, which will be used later to calculate the integrals on the Riemann surface:

$$\lim_{\varepsilon \to 0}\mathcal{W}(a \pm \varepsilon, s; a, b) = \pm\sqrt{b^2 - s^2} \qquad (s \leq b),$$

$$\mathcal{W}(r, s=0; a, b) = \text{sign}(r-a)\sqrt{(r-a)^2 + b^2}, \qquad (2.152)$$

$$\mathcal{W}(r, -s; a, b) = \overline{\mathcal{W}(r, s; a, b)}.$$

For the fundamental root, this finally leads to

$$W(r, -s; \mu, x, y) = \overline{W(r, s; \mu, x, y)},$$

$$W(r = \xi_i - \varepsilon, s; \mu, x, y) = -W(r = \xi_i + \varepsilon, s; \mu, x, y)$$

$$(i \in \{1,2,3\}, |s| < \xi_i, \varepsilon \text{ sufficiently small}),$$

$$W(r, s=0; \mu, x, y) = \text{sign}\left[(r-\xi_1)(r-\xi_2)(r-\xi_3)\right]$$

$$\times \mu\sqrt{(r-\xi_1)^2 + \eta_1^2}\sqrt{(r-\xi_2)^2 + \eta_2^2}\sqrt{(r-\xi_3)^2 + \eta_3^2}. \qquad (2.153)$$

---

[9] The roots that appear here are roots of positive, real quantities and must always be taken to have positive sign.

In the following formulae, (i,j,k) is any permutation of $(1,2,3)$. With

$$\begin{aligned}G_R(s;a_i) &:= \mathcal{W}_R(\xi_i,s;\xi_j,\eta_j)\mathcal{W}_R(\xi_i,s;\xi_k,\eta_k) \\ &\quad - \mathcal{W}_I(\xi_i,s;\xi_j,\eta_j)\mathcal{W}_I(\xi_i,s;\xi_k,\eta_k), \\ G_I(s;a_i) &:= \mathcal{W}_R(\xi_i,s;\xi_j,\eta_j)\mathcal{W}_I(\xi_i,s;\xi_k,\eta_k) \\ &\quad + \mathcal{W}_I(\xi_i,s;\xi_j,\eta_j)\mathcal{W}_R(\xi_i,s;\xi_k,\eta_k),\end{aligned} \qquad (2.154)$$

one finds

$$\begin{aligned}W_\pm^{a_i}(s;\mu,x,y) &:= \lim_{\varepsilon \to 0} W(\xi_i \pm \varepsilon, s; \mu, x, y) \\ &= \pm\mu\sqrt{\eta_i^2 - s^2}\,\mathcal{W}(\xi_i,s;\xi_j,\eta_j)\mathcal{W}(\xi_i,s;\xi_k,\eta_k) \qquad (2.155) \\ &= \pm\mu\sqrt{\eta_i^2 - s^2}\,[G_R(s;a_i) + \mathrm{i}\,G_I(s;a_i)]\end{aligned}$$

for the fundamental root along the cut $V(\xi_i, \eta_i)$. For the calculation of the normalizing coefficients $\alpha_0, \alpha_1, \beta_0, \beta_1, \gamma_0, \gamma_1$ and the quantities B_{11}, B_{12}, B_{22} the following notation is introduced:

$$\mathcal{A}_i^n(\mu, x, y) := \oint_{a_i} \frac{X^n\,\mathrm{d}X}{W(X)} = 2\mathrm{i} \int_{-\eta_i}^{\eta_i} \frac{(\xi_i + \mathrm{i}s)^n\,\mathrm{d}s}{W_+^{a_i}(s;\mu,x,y)}. \qquad (2.156)$$

Consider, for example, $\mathcal{A}_i^0$:

$$\begin{aligned}\mathcal{A}_i^0(\mu,x,y) &= \frac{2\mathrm{i}}{\mu} \int_{-\eta_i}^{\eta_i} \frac{G_R(s;a_i) - \mathrm{i}\,G_I(s;a_i)}{G_R^2(s;a_i) + G_I^2(s;a_i)} \frac{\mathrm{d}s}{\sqrt{\eta_i^2 - s^2}} \\ &= \frac{2}{\mu} \int_{-\eta_i}^{\eta_i} \frac{G_I(s;a_i)}{G_R^2(s;a_i) + G_I^2(s;a_i)} \frac{\mathrm{d}s}{\sqrt{\eta_i^2 - s^2}} \qquad (2.157) \\ &\quad + \frac{2\mathrm{i}}{\mu} \int_{-\eta_i}^{\eta_i} \frac{G_R(s;a_i)}{G_R^2(s;a_i) + G_I^2(s;a_i)} \frac{\mathrm{d}s}{\sqrt{\eta_i^2 - s^2}}.\end{aligned}$$

Due to the symmetry properties of $G_I(s;a_i)$, the real part of the equation above vanishes. Nevertheless, for the calculation of B_{11}, B_{12} and B_{22}, it is useful to use

2.3 The rigidly rotating disc of dust

the corresponding integrals over the interval $[0, \eta_i]$. Therefore the integrals

$$\mathcal{L}_i^0(\mu, x, y) := \frac{1}{\mu} \int_0^{\eta_i} \frac{G_R(s; a_i)}{G_R^2(s; a_i) + G_I^2(s; a_i)} \frac{ds}{\sqrt{\eta_i^2 - s^2}}, \tag{2.158a}$$

$$\mathcal{L}_i^1(\mu, x, y) := \frac{1}{\mu} \int_0^{\eta_i} \frac{\xi_i G_R(s; a_i) + s G_I(s; a_i)}{G_R^2(s; a_i) + G_I^2(s; a_i)} \frac{ds}{\sqrt{\eta_i^2 - s^2}}, \tag{2.158b}$$

$$\mathcal{L}_i^2(\mu, x, y) := \frac{1}{\mu} \int_0^{\eta_i} \frac{(\xi_i^2 - s^2) G_R(s; a_i) + 2\xi_i s G_I(s; a_i)}{G_R^2(s; a_i) + G_I^2(s; a_i)} \frac{ds}{\sqrt{\eta_i^2 - s^2}}, \tag{2.158c}$$

$$\mathcal{I}_i^0(\mu, x, y) := \frac{1}{\mu} \int_0^{\eta_i} \frac{G_I(s; a_i)}{G_R^2(s; a_i) + G_I^2(s; a_i)} \frac{ds}{\sqrt{\eta_i^2 - s^2}}, \tag{2.158d}$$

$$\mathcal{I}_i^1(\mu, x, y) := \frac{1}{\mu} \int_0^{\eta_i} \frac{\xi_i G_I(s; a_i) - s G_R(s; a_i)}{G_R^2(s; a_i) + G_I^2(s; a_i)} \frac{ds}{\sqrt{\eta_i^2 - s^2}}, \tag{2.158e}$$

$$\mathcal{I}_i^2(\mu, x, y) := \frac{1}{\mu} \int_0^{\eta_i} \frac{(\xi_i^2 - s^2) G_I(s; a_i) - 2\xi_i s G_R(s; a_i)}{G_R^2(s; a_i) + G_I^2(s; a_i)} \frac{ds}{\sqrt{\eta_i^2 - s^2}} \tag{2.158f}$$

are defined and hence

$$\mathcal{A}_i^n(\mu, x, y) = 4\mathrm{i}\, \mathcal{L}_i^n(\mu, x, y). \tag{2.159}$$

For the calculation of the moduli B_{mn} and the quantities C_1, C_2, integrals of the type

$$\mathcal{K}_{c,d}^n(\mu, x, y) := \int_c^d \frac{r^n dr}{W(r, s = 0; \mu, x, y)} \quad (n = 0, 1) \tag{2.160}$$

are also used. These integrals, which run along parts of the real axis, can be computed using the last equation of (2.153).

Due to Equations (2.132), (2.133) and (2.159) we have

$$\alpha_0 \mathcal{L}_1^0 + \alpha_1 \mathcal{L}_1^1 = \frac{\pi}{4}, \qquad \beta_0 \mathcal{L}_1^0 + \beta_1 \mathcal{L}_1^1 = 0, \tag{2.161}$$

$$\alpha_0 \mathcal{L}_2^0 + \alpha_1 \mathcal{L}_2^1 = 0, \qquad \beta_0 \mathcal{L}_2^0 + \beta_1 \mathcal{L}_2^1 = \frac{\pi}{4}. \tag{2.162}$$

A consequence of these relations and of

$$\oint_{a_1} d\omega_n + \oint_{a_2} d\omega_n + \oint_{a_3} d\omega_n = 0 \qquad (n=1,2) \qquad (2.163)$$

is

$$\alpha_0 \mathcal{L}_3^0 + \alpha_1 \mathcal{L}_3^1 = -\frac{\pi}{4} \quad \text{and} \quad \beta_0 \mathcal{L}_3^0 + \beta_1 \mathcal{L}_3^1 = -\frac{\pi}{4}, \qquad (2.164)$$

which will be used in the context of the calculation of B_{mn} and C_m.

Determination of $\alpha_0, \alpha_1, \beta_0, \beta_1$: Due to (2.161) and (2.162), we have to solve the two linear systems

$$\begin{pmatrix} \mathcal{L}_1^0 & \mathcal{L}_1^1 \\ \mathcal{L}_2^0 & \mathcal{L}_2^1 \end{pmatrix} \begin{pmatrix} \alpha_0 \\ \alpha_1 \end{pmatrix} = \begin{pmatrix} \pi/4 \\ 0 \end{pmatrix}, \qquad (2.165)$$

$$\begin{pmatrix} \mathcal{L}_1^0 & \mathcal{L}_1^1 \\ \mathcal{L}_2^0 & \mathcal{L}_2^1 \end{pmatrix} \begin{pmatrix} \beta_0 \\ \beta_1 \end{pmatrix} = \begin{pmatrix} 0 \\ \pi/4 \end{pmatrix}. \qquad (2.166)$$

The solutions are

$$\alpha_0 = \frac{\pi}{4D_a}\mathcal{L}_2^1, \alpha_1 = -\frac{\pi}{4D_a}\mathcal{L}_2^0 \quad \text{and} \quad \beta_0 = -\frac{\pi}{4D_a}\mathcal{L}_1^1, \beta_1 = \frac{\pi}{4D_a}\mathcal{L}_1^0, \qquad (2.167)$$

where $D_a := \mathcal{L}_1^0 \mathcal{L}_2^1 - \mathcal{L}_2^0 \mathcal{L}_1^1$.

Determination of γ_1 and γ_2: The definition (2.134) and the condition (2.135) lead to the system

$$\begin{pmatrix} \mathcal{L}_1^0 & \mathcal{L}_1^1 \\ \mathcal{L}_2^0 & \mathcal{L}_2^1 \end{pmatrix} \begin{pmatrix} \gamma_0 \\ \gamma_1 \end{pmatrix} = -\mu \begin{pmatrix} \mathcal{L}_1^2 \\ \mathcal{L}_2^2 \end{pmatrix} \qquad (2.168)$$

with the solutions

$$\gamma_0 = -\mu \frac{\mathcal{L}_1^2 \mathcal{L}_2^1 - \mathcal{L}_2^2 \mathcal{L}_1^1}{D_a}, \qquad \gamma_1 = -\mu \frac{\mathcal{L}_1^0 \mathcal{L}_2^2 - \mathcal{L}_2^0 \mathcal{L}_1^2}{D_a}. \qquad (2.169)$$

Determination of B_{11}, B_{22}, B_{12} and C_1, C_2: It turns out that for the calculation of these quantities, two cases have to be discussed separately, namely $y > \Re X_2$ (corresponding to the cut $V(\xi_3, \eta_3)$ being to the right of $V(\xi_2, \eta_2)$, see Fig. 2.7) and $0 < y < \Re X_2$ (corresponding to $V(\xi_3, \eta_3)$ being to the left of $V(\xi_2, \eta_2)$, see Fig. 2.8). In both cases the path of integration is divided into suitable parts. We use the following representation for B_{mn} and C_m ($m, n = 1, 2$)

$$B_{mn} = \oint_{b_m} d\omega_n = 2 \sum_l \int_{b_{m(l)}} d\omega_n, \quad C_n = \int_{+\infty}^{P} d\omega_n = \sum_l \int_{c_{(l)}} d\omega_n. \qquad (2.170)$$

2.3 The rigidly rotating disc of dust

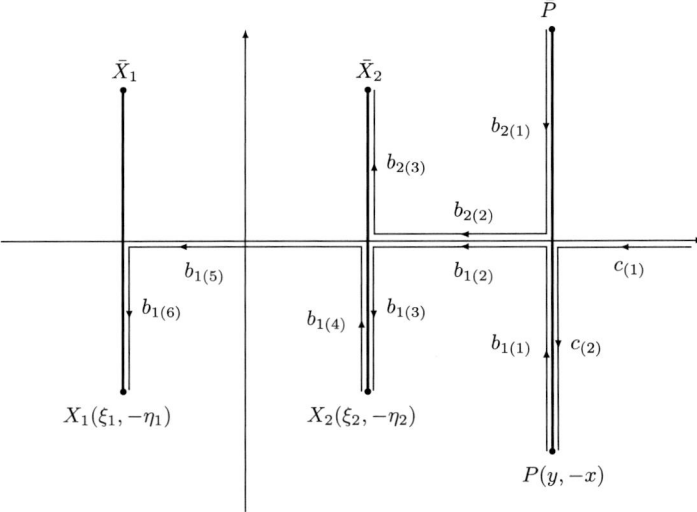

Fig. 2.7. Complex plane with cuts $V(\xi_1, \eta_1)$, $V(\xi_2, \eta_2)$ and $V(\xi_3 = y, \eta_3 = x)$ for the case $y > \Re X_2$. This case corresponds to the Riemann surface shown in Fig. 2.6.

The structure $B_{mn} = 2\{\ldots\}$ is due to the fact that on the Riemann surface (see Fig. 2.6), one also has to integrate the corresponding return path in the lower sheet for calculating the moduli.

The case $y > \Re X_2$:

From Fig. 2.7, one can read off the corresponding B_{mn}:

$$B_{11} = 2\{\alpha_0(\mathcal{I}_3^0 + \mathcal{K}_{y,\xi_2}^0 + 2\mathcal{I}_2^0 + \mathcal{K}_{\xi_2,\xi_1}^0 + \mathcal{I}_1^0)$$
$$+ \alpha_1(\mathcal{I}_3^1 + \mathcal{K}_{y,\xi_2}^1 + 2\mathcal{I}_2^1 + \mathcal{K}_{\xi_2,\xi_1}^1 + \mathcal{I}_1^1)\},$$
$$B_{22} = 2\{\beta_0(\mathcal{I}_3^0 + \mathcal{K}_{y,\xi_2}^0 + \mathcal{I}_2^0) + \beta_1(\mathcal{I}_3^1 + \mathcal{K}_{y,\xi_2}^1 + \mathcal{I}_2^1)\},$$
$$B_{12} = 2\Big\{\beta_0(\mathcal{I}_3^0 + \mathcal{K}_{y,\xi_2}^0 + 2\mathcal{I}_2^0 + \mathcal{K}_{\xi_2,\xi_1}^0 + \mathcal{I}_1^0) \qquad (2.171)$$
$$+ \beta_1(\mathcal{I}_3^1 + \mathcal{K}_{y,\xi_2}^1 + 2\mathcal{I}_2^1 + \mathcal{K}_{\xi_2,\xi_1}^1 + \mathcal{I}_1^1) - \frac{i\pi}{4}\Big\},$$
$$B_{21} = 2\Big\{\alpha_0(\mathcal{I}_3^0 + \mathcal{K}_{y,\xi_2}^0 + \mathcal{I}_2^0) + \alpha_1(\mathcal{I}_3^1 + \mathcal{K}_{y,\xi_2}^1 + \mathcal{I}_2^1) - \frac{i\pi}{4}\Big\},$$

and C_1, C_2 are given by

$$C_1 = \alpha_0(\mathcal{K}_{\infty,y}^0 + \mathcal{I}_3^0) + \alpha_1(\mathcal{K}_{\infty,y}^1 + \mathcal{I}_3^1) + \frac{i\pi}{4},$$
$$C_2 = \beta_0(\mathcal{K}_{\infty,y}^0 + \mathcal{I}_3^0) + \beta_1(\mathcal{K}_{\infty,y}^1 + \mathcal{I}_3^1) + \frac{i\pi}{4}. \qquad (2.172)$$

66   *Analytical treatment of limiting cases*

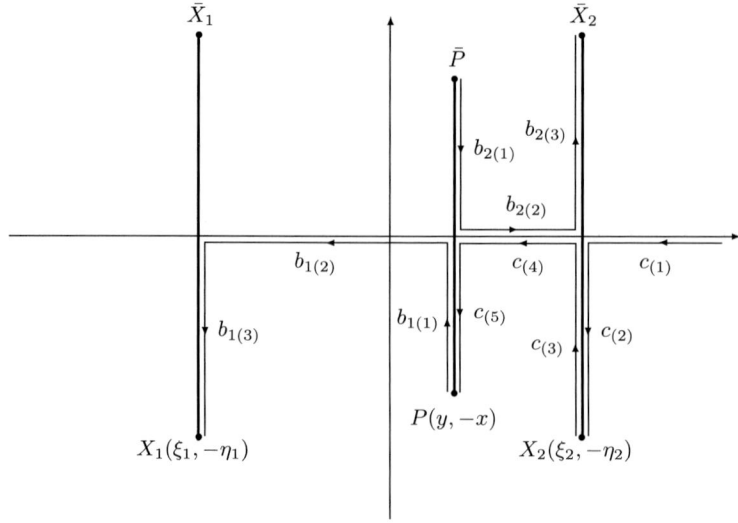

Fig. 2.8. Complex plane with cuts $V(\xi_1, \eta_1)$, $V(\xi_2, \eta_2)$ and $V(\xi_3 = y, \eta_3 = x)$ for the case $0 < y < \Re X_2$.

The case $0 < y < \Re X_2$:
Analogously, from Fig. 2.8 one finds B_{mn}:

$$B_{11} = 2\{\alpha_0(\mathcal{I}_3^0 + \mathcal{K}_{y,\xi_1}^0 + \mathcal{I}_1^0) + \alpha_1(\mathcal{I}_3^1 + \mathcal{K}_{y,\xi_1}^1 + \mathcal{I}_1^1)\},$$
$$B_{22} = 2\{\beta_0(-\mathcal{I}_3^0 + \mathcal{K}_{y,\xi_2}^0 - \mathcal{I}_2^0) + \beta_1(-\mathcal{I}_3^1 + \mathcal{K}_{y,\xi_2}^1 - \mathcal{I}_2^1)\},$$
$$B_{12} = 2\left\{\beta_0(\mathcal{I}_3^0 + \mathcal{K}_{y,\xi_1}^0 + \mathcal{I}_1^0) + \beta_1(\mathcal{I}_3^1 + \mathcal{K}_{y,\xi_1}^1 + \mathcal{I}_1^1) + \frac{i\pi}{4}\right\},$$
$$B_{21} = 2\left\{\alpha_0(-\mathcal{I}_3^0 + \mathcal{K}_{y,\xi_2}^0 - \mathcal{I}_2^0) + \alpha_1(-\mathcal{I}_3^1 + \mathcal{K}_{y,\xi_2}^1 - \mathcal{I}_2^1) + \frac{i\pi}{4}\right\}, \quad (2.173)$$

and C_1, C_2 are given by

$$\begin{aligned} C_1 &= \alpha_0(\mathcal{K}_{\infty,\xi_2}^0 + 2\mathcal{I}_2^0 + \mathcal{K}_{\xi_2,y}^0 + \mathcal{I}_3^0) \\ &\quad + \alpha_1(\mathcal{K}_{\infty,\xi_2}^1 + 2\mathcal{I}_2^1 + \mathcal{K}_{\xi_2,y}^1 + \mathcal{I}_3^1) + \frac{i\pi}{4}, \\ C_2 &= \beta_0(\mathcal{K}_{\infty,\xi_2}^0 + 2\mathcal{I}_2^0 + \mathcal{K}_{\xi_2,y}^0 + \mathcal{I}_3^0) \\ &\quad + \beta_1(\mathcal{K}_{\infty,\xi_2}^1 + 2\mathcal{I}_2^1 + \mathcal{K}_{\xi_2,y}^1 + \mathcal{I}_3^1) - \frac{i\pi}{4}. \end{aligned} \quad (2.174)$$

The imaginary parts of the equations follow directly from (2.161), (2.162) and (2.164). The symbol ∞ in $\mathcal{K}_{\infty,y}^n$ stands for '+ real infinity'. Note that $B_{21} = B_{12}$ holds in general, see e.g. Krazer (1903). Since different expressions for each of

2.3 The rigidly rotating disc of dust

these quantities can be found in (2.171) and (2.173), one can derive useful identities and one has a further test for verifying a numerical evaluation of these integrals.

Using symmetry properties, the functions u, v and w (2.129) are found to be[10]

$$u(\mu, x, y) = \frac{2}{\pi} \int_0^1 \frac{\ln[\sqrt{1 + \mu^2(1-s^2)^2} + \mu(1-s^2)]}{\sqrt{1 + \mu^2(1-s^2)^2}} \frac{\mathcal{W}_R(0, s; y, x)}{\mathcal{W}_R^2 + \mathcal{W}_I^2} ds,$$

$$v(\mu, x, y) = \frac{2}{\pi} \int_0^1 \frac{\ln[\sqrt{1 + \mu^2(1-s^2)^2} + \mu(1-s^2)]}{\sqrt{1 + \mu^2(1-s^2)^2}} \frac{s\mathcal{W}_I(0, s; y, x)}{\mathcal{W}_R^2 + \mathcal{W}_I^2} ds,$$

$$w(\mu, x, y) = -\frac{2}{\pi} \int_0^1 \frac{\ln[\sqrt{1 + \mu^2(1-s^2)^2} + \mu(1-s^2)]}{\sqrt{1 + \mu^2(1-s^2)^2}} \frac{s^2 \mathcal{W}_R(0, s; y, x)}{\mathcal{W}_R^2 + \mathcal{W}_I^2} ds.$$

(2.175)

Transformation of the theta formula: Now, using relations between Rosenhain's theta functions, the solution for the Ernst potential is transformed in such a way that all the arguments become real. First we write down the relation

$$\vartheta_{4,4}(x_1, x_2; B_{11}, B_{22}, B_{12}) =$$
$$\vartheta_{3,3}(x_1 + x_2, x_1 - x_2; B_{11} + 2B_{12} + B_{22}, B_{11} - 2B_{12} + B_{22}, B_{11} - B_{22})$$
$$- \vartheta_{2,2}(x_1 + x_2, x_1 - x_2; B_{11} + 2B_{12} + B_{22}, B_{11} - 2B_{12} + B_{22}, B_{11} - B_{22}),$$
(2.176)

which can be deduced from the definitions (A2.10). It will be convenient to introduce the following combinations of the arguments:

$$L := \exp\{-(\gamma_0 u + \gamma_1 v + \mu w)\}, \quad S := (\alpha_0 + \beta_0)u + (\alpha_1 + \beta_1)v, \quad (2.177)$$

$$T := B_{11} + B_{22} + 2\Re B_{12}, \quad A := (\alpha_0 - \beta_0)u + (\alpha_1 - \beta_1)v, \quad (2.178)$$

$$B := B_{11} + B_{22} - 2\Re B_{12}, \quad C := \Re(C_1 + C_2), \quad (2.179)$$

$$R := B_{11} - B_{22}, \quad D := \Re(C_1 - C_2). \quad (2.180)$$

The previous results lead immediately to the equations (upper sign corresponds to $y > \Re X_2$ and lower sign to $0 < y < \Re X_2$)

$$B_{12} = \Re B_{12} \mp i\frac{\pi}{2}, \quad C_1 = \Re C_1 + i\frac{\pi}{4}, \quad C_2 = \Re C_2 \pm i\frac{\pi}{4},$$
$$B_{11} + B_{22} + 2B_{12} = T \mp i\pi, \quad B_{11} + B_{22} - 2B_{12} = B \pm i\pi,$$
(2.181)

---

[10] The arguments of the denominator $\mathcal{W}_R^2 + \mathcal{W}_I^2$ are always the same as for the corresponding numerator $\mathcal{W}_{R/I}$.

and hence

$$y > \Re X_2 : \quad C_1 + C_2 = \Re(C_1 + C_2) + i\frac{\pi}{2}, \quad C_1 - C_2 = \Re(C_1 - C_2),$$

$$0 < y < \Re X_2 : \quad C_1 + C_2 = \Re(C_1 + C_2), \quad C_1 - C_2 = \Re(C_1 - C_2) + i\frac{\pi}{2}. \tag{2.182}$$

Applying (2.176) to (2.127) and using the properties (A2.11), (A2.12) leads to the results :

Case $\quad y > \Re X_2$:

$$f = L\frac{\vartheta_{3,4}(S-C, A-D; T, B, R) + i\vartheta_{1,2}(S-C, A-D; T, B, R)}{\vartheta_{3,4}(S+C, A+D; T, B, R) - i\vartheta_{1,2}(S+C, A+D; T, B, R)}. \tag{2.183}$$

Case $\quad 0 < y < \Re X_2$:

$$f = L\frac{\vartheta_{4,3}(S-C, A-D; T, B, R) + i\vartheta_{2,1}(S-C, A-D; T, B, R)}{\vartheta_{4,3}(S+C, A+D; T, B, R) - i\vartheta_{2,1}(S+C, A+D; T, B, R)}. \tag{2.184}$$

Finally, with $\vartheta_{n,k}^{\pm} := \vartheta_{n,k}(S \pm C, A \pm D; T, B, R)$, one finds for e^{2U} and b :

Case $\quad y > \Re X_2$:

$$e^{2U} = L\frac{\vartheta_{3,4}^{-}\vartheta_{3,4}^{+} - \vartheta_{1,2}^{-}\vartheta_{1,2}^{+}}{\left[\vartheta_{3,4}^{+}\right]^2 + \left[\vartheta_{1,2}^{+}\right]^2}, \quad b = L\frac{\vartheta_{3,4}^{-}\vartheta_{1,2}^{+} + \vartheta_{3,4}^{+}\vartheta_{1,2}^{-}}{\left[\vartheta_{3,4}^{+}\right]^2 + \left[\vartheta_{1,2}^{+}\right]^2}. \tag{2.185}$$

Case $\quad 0 < y < \Re X_2$:

$$e^{2U} = L\frac{\vartheta_{4,3}^{-}\vartheta_{4,3}^{+} - \vartheta_{2,1}^{-}\vartheta_{2,1}^{+}}{\left[\vartheta_{4,3}^{+}\right]^2 + \left[\vartheta_{2,1}^{+}\right]^2}, \quad b = L\frac{\vartheta_{4,3}^{-}\vartheta_{2,1}^{+} + \vartheta_{4,3}^{+}\vartheta_{2,1}^{-}}{\left[\vartheta_{4,3}^{+}\right]^2 + \left[\vartheta_{2,1}^{+}\right]^2}. \tag{2.186}$$

Applications: The arguments (2.177)–(2.180) of Equations (2.186), (2.185) are given by (2.167), (2.169), [(2.171) and (2.172) for $y > \Re X_2$] or [(2.173) and (2.174) for $0 < y < \Re X_2$] and (2.175). These quantities can be calculated as shown above with the three types of integrals (2.158a)–(2.158c), (2.158d)–(2.158e) and (2.160). These integrals themselves can be numerically evaluated very easily. Whether or not the y-coordinate approaches $\Re X_2$ from the left or the right, (2.186) and (2.185) converge to the same result.[11] In this region, a little more attention must be paid to

---

[11] The limit $(\xi_3 = y, \eta_3 = x) \to (\xi_2, \eta_2)$ can be calculated analytically.

2.3 The rigidly rotating disc of dust

the numerical evaluation of the integrals. The theta functions, on the other hand, are easy to handle numerically, since the rapidly converging series make it possible to replace the infinite sums (A2.1) and (A2.10) by finite sums of the type $\sum_{n=-j}^{j}$, with, say $j = 10$.

Equations (2.186) and (2.185) and the formulae for the related arguments are also very well suited to deriving simpler formulae for the special cases of the plane of symmetry ($y = 0$) and the axis ($x = 0$). The simplest form of the Ernst potential and the whole metric is to be found within the disc ($y = 0, x \leq 1$) (Kleinwächter 2000). How these formulae can be derived will be explained in some detail for the latter case. For the other two cases, we just list the results.

Disc metric

In this case, one has to consider $y = 0^+, x \leq 1$ and thus to use the formulae for the case $0 < y < \Re X_2$. It turns out that

$$
\begin{aligned}
&\beta_0 = \alpha_0, & &C = 2\Re(C_1), & &T = 2\Re(B_{11} + B_{12}), \\
&\beta_1 = -\alpha_1, & &D = 0, & &B = 2\Re(B_{11} - B_{12}), \\
&\gamma_1 = 0, & &S = 2\alpha_0 u, & &R = 0, \\
&B_{11} = B_{22}, & &A = 2\alpha_1 v, & &L = \exp\{-(\gamma_0 u + \mu w)\}.
\end{aligned} \tag{2.187}
$$

Using the general identity $\vartheta_{n,k}(x_1, x_2; a, b, c = 0) = \vartheta_n(x_1; a)\vartheta_k(x_2; b)$, one finds

$$
f = L \frac{\vartheta_4(S - C; T)\vartheta_3(A; B) + i\vartheta_2(S - C; T)\vartheta_1(A; B)}{\vartheta_4(S + C; T)\vartheta_3(A; B) - i\vartheta_2(S + C; T)\vartheta_1(A; B)},
$$

$$
f = L \frac{\vartheta_2(S - C; T)}{\vartheta_2(S + C; T)} \frac{\frac{\vartheta_4(S-C;T)}{\vartheta_2(S-C;T)} + i \frac{\vartheta_1(A;B)}{\vartheta_3(A;B)}}{\frac{\vartheta_4(S+C;T)}{\vartheta_2(S+C;T)} - i \frac{\vartheta_1(A;B)}{\vartheta_3(A;B)}}. \tag{2.188}
$$

For convenience we introduce a new variable

$$
\tilde{\mu} := \mu(1 - x^2). \tag{2.189}
$$

Within the disc, the functions u, v and w given by (2.175) in this new variable simplify to

$$
\begin{aligned}
u(\mu, \tilde{\mu}) &= -\frac{1}{\pi} \int_{\tilde{\mu}}^{\mu} \frac{\ln(\sqrt{1+t^2}+t)}{\sqrt{1+t^2}} \frac{dt}{\sqrt{\mu-t}\sqrt{t-\tilde{\mu}}}, \\
\sqrt{\tilde{\mu}} v(\mu, \tilde{\mu}) &= \frac{1}{\pi} \int_{0}^{\tilde{\mu}} \frac{\ln(\sqrt{1+t^2}+t)}{\sqrt{1+t^2}} \frac{dt}{\sqrt{\tilde{\mu}-t}}, \\
\mu w(\mu, \tilde{\mu}) &= \frac{1}{\pi} \int_{\tilde{\mu}}^{\mu} \frac{\ln(\sqrt{1+t^2}+t)}{\sqrt{1+t^2}} \frac{\sqrt{\mu-t}}{\sqrt{t-\tilde{\mu}}} dt.
\end{aligned} \tag{2.190}
$$

Moreover, one can express the functions u and w through elliptic functions and elliptic theta functions

$$u(\mu, \tilde{\mu}) = -\frac{1}{\sqrt[4]{(1+\mu^2)(1+\tilde{\mu}^2)}} F(\varphi, l'),$$

$$\mu w(\mu, \tilde{\mu}) = -\ln\left(\frac{\mu + \tilde{\mu}}{2} + \sqrt{1+\tilde{\mu}^2}\right)$$

$$- F(\varphi, l') \left[\frac{\sqrt{1+\tilde{\mu}^2} - (\mu - \tilde{\mu})}{\sqrt[4]{(1+\mu^2)(1+\tilde{\mu}^2)}} - \frac{\pi}{K(l)} \Lambda_0(\psi, l)\right] \quad (2.191)$$

$$+ \ln \frac{\vartheta_3\left(\frac{\pi}{2K(l)}\left[F(\varphi, l') - F(\psi, l')\right]; -\pi \frac{K(l')}{K(l)}\right)}{\vartheta_3\left(\frac{\pi}{2K(l)}\left[F(\varphi, l') + F(\psi, l')\right]; -\pi \frac{K(l')}{K(l)}\right)}.$$

The main arguments are

$$\varphi(\mu, \tilde{\mu}) = \arcsin\sqrt{\frac{\sqrt{(1+\mu^2)(1+\tilde{\mu}^2)} + \mu\tilde{\mu} - 1}{\sqrt{(1+\mu^2)(1+\tilde{\mu}^2)} + \mu\tilde{\mu} + 1}},$$

$$\psi(\mu, \tilde{\mu}) = \arcsin\sqrt{\frac{\sqrt{(1+\mu^2)(1+\tilde{\mu}^2)} + \tilde{\mu}^2 + 1 - \sqrt{1+\tilde{\mu}^2}(\mu - \tilde{\mu})}{\sqrt{(1+\mu^2)(1+\tilde{\mu}^2)} + \mu\tilde{\mu} + 1}},$$

(2.192)

the modulus l is given by

$$l(\mu, \tilde{\mu}) = \sqrt{\frac{1}{2}\left(1 - \frac{1 + \mu\tilde{\mu}}{\sqrt{1+\mu^2}\sqrt{1+\tilde{\mu}^2}}\right)} \quad (2.193)$$

and the complementary modulus l' is

$$l' \equiv \sqrt{1 - l^2}. \quad (2.194)$$

Furthermore the moduli T and B of the Jacobian theta functions are

$$T(\mu, \tilde{\mu}) = -\pi \frac{K(l')}{K(l)}, \qquad B(\tilde{\mu}) = -\pi \frac{K(\tilde{h}')}{K(\tilde{h})}, \quad (2.195)$$

with the additional moduli $\tilde{h}$ and $\tilde{h}'$

$$\tilde{h} \equiv h(\tilde{\mu}) \equiv \sqrt{\frac{1}{2}\left(1 + \frac{\tilde{\mu}}{\sqrt{1+\tilde{\mu}^2}}\right)}, \qquad \tilde{h}' \equiv \sqrt{1 - \tilde{h}^2}. \quad (2.196)$$

2.3 The rigidly rotating disc of dust

We make use of the following notation for functions that are closely related to v as given in (2.190),

$$I(\mu) := \frac{1}{\pi}\int_0^\mu \frac{\ln(\sqrt{1+t^2}+t)}{\sqrt{1+t^2}}\frac{dt}{\sqrt{\mu-t}}, \qquad \hat{I}(\mu) := \sqrt[4]{1+\mu^2}\,I(\mu). \qquad (2.197)$$

All the other arguments in (2.188) follow:

$$\alpha_0 = \frac{\pi}{4\mathrm{K}(l)}\sqrt[4]{(1+\mu^2)(1+\tilde{\mu}^2)},$$

$$\alpha_1 = -\frac{\pi}{4\mathrm{K}(\tilde{h})}\sqrt[4]{1+\tilde{\mu}^2}\sqrt{\mu},$$

$$\gamma_0 = -\frac{\pi}{\mathrm{K}(l)}\sqrt[4]{(1+\mu^2)(1+\tilde{\mu}^2)}\left[1+\frac{1}{\pi}\mathrm{K}(l)\frac{\sqrt{1+\tilde{\mu}^2}-(\mu-\tilde{\mu})}{\sqrt[4]{(1+\mu^2)(1+\tilde{\mu}^2)}}-\Lambda_0(\psi,l)\right],$$

$$S = 2\alpha_0 u(\mu,\tilde{\mu}) = -\frac{\pi}{2\mathrm{K}(l)}\mathrm{F}(\varphi,l'),$$

$$A = 2\alpha_1 v(\mu,\tilde{\mu}) = -\frac{\pi}{2\mathrm{K}(\tilde{h})}\sqrt[4]{(1+\tilde{\mu}^2)}I(\tilde{\mu}) = -\frac{\pi}{2\mathrm{K}(\tilde{h})}\hat{I}(\tilde{\mu}),$$

$$C = \frac{\pi}{2\mathrm{K}(l)}\left[\mathrm{K}(l')-\mathrm{F}(\psi,l')\right]$$

and

$$L = \left(\sqrt{1+\tilde{\mu}^2}+\frac{\mu+\tilde{\mu}}{2}\right)\exp\left[-\frac{\pi}{\mathrm{K}(l)}\mathrm{F}(\varphi,l')\right]$$

$$\times \frac{\vartheta_3\left(\frac{\pi}{2\mathrm{K}(l)}[\mathrm{F}(\varphi,l')+\mathrm{F}(\psi,l')];T\right)}{\vartheta_3\left(\frac{\pi}{2\mathrm{K}(l)}[\mathrm{F}(\varphi,l')-\mathrm{F}(\psi,l')];T\right)}. \qquad (2.198)$$

Putting all these formulae into (2.188) and using the relations between the Jacobian theta functions and Jacobian elliptic functions (A2.20)–(A2.22), one finally finds

$$e^{2U}(\mu,\tilde{\mu}) = \frac{h'(\tilde{\mu})}{h(\tilde{\mu})}\mathrm{cn}^2\left[\hat{I}(\tilde{\mu}),h'(\tilde{\mu})\right]-\frac{\mu-\tilde{\mu}}{2}, \qquad (2.199)$$

$$b(\mu,\tilde{\mu}) = -\frac{1}{h(\tilde{\mu})}\mathrm{sn}\left[\hat{I}(\tilde{\mu}),h'(\tilde{\mu})\right]\mathrm{dn}\left[\hat{I}(\tilde{\mu}),h'(\tilde{\mu})\right]. \qquad (2.200)$$

Specialization to the origin ($x = 0$ is equivalent to $\tilde{\mu} = \mu$) leads to

$$e^{2V_0}(\mu) \equiv e^{2U}(\mu,\mu) = \frac{h'(\mu)}{h(\mu)}\mathrm{cn}^2\left[\hat{I}(\mu),h'(\mu)\right], \qquad (2.201)$$

$$b_0(\mu) \equiv b(\mu,\mu) = -\frac{1}{h(\mu)}\mathrm{sn}\left[\hat{I}(\mu),h'(\mu)\right]\mathrm{dn}\left[\hat{I}(\mu),h'(\mu)\right]. \qquad (2.202)$$

From (2.199), (2.200), (2.201) and (2.202) one immediately finds the scaling property (Neugebauer and Meinel 1994)

$$e^{2U}(\mu, x) = e^{2V_0}\left[\mu(1-x^2)\right] - \frac{\mu}{2}x^2, \quad b(\mu, x) = b_0\left[\mu(1-x^2)\right], \quad (2.203)$$

now reusing the original variables μ and x.

The parameter function $\Omega\varrho_0$ follows from (2.79):

$$\Omega\varrho_0(\mu) = \frac{1}{2}\sqrt{1 - \frac{h'^2(\mu)}{h^2(\mu)}} \operatorname{cn}\left[\hat{I}(\mu), h'(\mu)\right]. \quad (2.204)$$

With (2.201), (2.202), (2.204) and using the general identities (A2.19) one can easily verify the parameter relation

$$e^{4V_0} + b_0^2 + 4\Omega^2\rho_0^2 = 1 \quad \text{(see (2.78) with: } f_0 = e^{2V_0} + i\,b_0). \quad (2.205)$$

The value $\mu_0 = 4.62966184\ldots$ (see (2.79)) is the smallest positive zero of $\operatorname{cn}\left[\hat{I}(\mu), h'(\mu)\right]$, i.e.

$$\operatorname{cn}\left[\hat{I}(\mu_0), h'(\mu_0)\right] = 0. \quad (2.206)$$

The monotonic function $V_0(\mu)$ varies over the range $0 \geq V_0 > -\infty$ when μ takes the values $0 \leq \mu < \mu_0$. The parameter functions e^{2V_0}, b_0 and $\Omega\varrho_0$ are plotted in Fig. 2.9.

In a similar manner as shown above for the Ernst potential, the general formulae for the metric functions Ωa (2.119) and e^{2k} (2.120) can be simplified within the

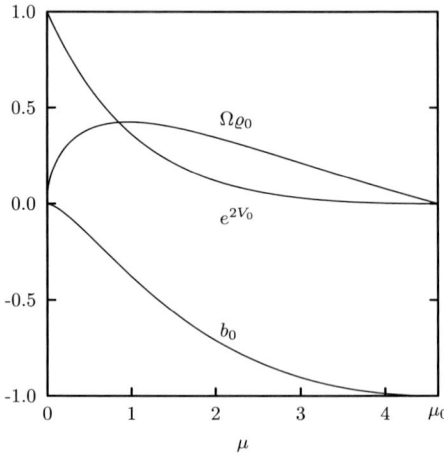

Fig. 2.9. The parameter functions e^{2V_0}, b_0 and $\Omega\varrho_0$.

2.3 The rigidly rotating disc of dust

disc. The result for $(1+\Omega a)e^{2U}$ is

$$(1+\Omega a)e^{2U} = \sqrt{\frac{h'}{h}}\sqrt{\frac{\tilde{h}'}{\tilde{h}}}\,\mathrm{cn}\left[\hat{I}(\mu),h'\right]\mathrm{cn}\left[\hat{I}(\tilde{\mu}),\tilde{h}'\right] = e^{V_0}(\mu)e^{V_0}(\tilde{\mu}). \quad (2.207)$$

Hence, for e^{2V} within the disc one gets

$$\begin{aligned} e^{2V} &= e^{2U}\left[(1+\Omega a)^2 - \Omega^2\varrho^2 e^{-4U}\right] \\ &= \frac{1}{e^{2U}}\left\{\left[(1+\Omega a)e^{2U}\right]^2 - (\Omega\varrho_0)^2 x^2\right\} \\ &= \frac{1}{e^{2U}}\left\{e^{2V_0}(\mu)e^{2V_0}(\tilde{\mu}) - \frac{\mu}{2}x^2 e^{2V_0}(\mu)\right\} \\ &= e^{2V_0}(\mu) \end{aligned} \quad (2.208)$$

thereby verifying that the boundary condition for the real part of the Ernst potential is indeed fulfilled, see (2.39). It turns out that three of the metric coefficients g_{ij} can be expressed with the parameter function

$$e^{V_0}(\mu) = \sqrt{\frac{h'(\mu)}{h(\mu)}}\,\mathrm{cn}\left[\hat{I}(\mu),h'(\mu)\right] \quad (2.209)$$

in an especially simple manner. The complete disc metric is given by[12]

$$\begin{aligned} \Omega^2 g_{\varphi\varphi}(\mu,x) &= -\left[e^{V_0}(\mu) - e^{V_0}(\tilde{\mu})\right]^2 + \frac{\mu x^2}{2}, \\ \Omega g_{\varphi t}(\mu,x) &= -e^{V_0}(\tilde{\mu})\left[e^{V_0}(\mu) - e^{V_0}(\tilde{\mu})\right] - \frac{\mu x^2}{2}, \\ g_{tt}(\mu,x) &= -e^{2V_0}(\tilde{\mu}) + \frac{\mu x^2}{2}, \\ g_{\varrho\varrho}(\mu,x) &= e^{2k_0}e^{\mathcal{F}}\frac{\pi K(h)}{K(\tilde{h})K(l)}\frac{\Theta_2^2(F(\varphi,l'),l)}{\mu-\tilde{\mu}}\frac{\Theta_2^2(\hat{I}(\tilde{\mu}),\tilde{h})}{\Theta_4^2(\hat{I}(\mu),h)}, \end{aligned} \quad (2.210)$$

with the function $\mathcal{F}$ defined by

$$\begin{aligned} \mathcal{F} &= \hat{I}^2(\mu)\frac{\Theta_4''(0,h)}{\Theta_4(0,h)} - \frac{1}{h'}\left\{F^2(\varphi,l')\left[l'\tilde{h}'\frac{\Theta_4''(0,l)}{\Theta_4(0,l)} - l\tilde{h}\frac{\Theta_2''(0,l)}{\Theta_2(0,l)}\right]\right. \\ &\quad + \hat{I}^2(\tilde{\mu})\left[l'\tilde{h}'\frac{\Theta_4''(0,\tilde{h})}{\Theta_4(0,\tilde{h})} - l\tilde{h}\frac{\Theta_2''(0,\tilde{h})}{\Theta_2(0,\tilde{h})}\right]\right\}. \end{aligned} \quad (2.211)$$

---

[12] The zero in the denominator of $g_{\varrho\varrho}$ for $x=0\,(\to\tilde{\mu}=\mu)$ is also a zero of the numerator and the metric coefficient remains regular.

74 *Analytical treatment of limiting cases*

One should keep in mind that $\tilde{\mu} = \mu(1 - x^2)$ and that the functions l, l' and h, h' are defined by (2.193), (2.196) and $\varphi, \hat{I}$ are defined by (2.192), (2.197). Furthermore $h \equiv h(\mu), \tilde{h} \equiv h(\tilde{\mu})$ and $h' \equiv h'(\mu), \tilde{h}' \equiv h'(\tilde{\mu})$. The theta functions Θ_2, Θ_4 are defined in Appendix 2 and Θ_2'', Θ_4'' denote second derivatives with respect to the main argument.

At the rim of the disc ($x = 1 \to \tilde{\mu} = 0$), we have, due to $e^{2V_0}(0) = 1$, a particularly simple form of $g_{\varphi\varphi}, g_{\varphi t}$ and g_{tt}:

$$\Omega^2 g_{\varphi\varphi}(\mu, 1) = -\left[1 - e^{V_0}(\mu)\right]^2 + \frac{\mu}{2},$$

$$\Omega g_{\varphi t}(\mu, 1) = 1 - e^{V_0}(\mu) - \frac{\mu}{2}, \qquad (2.212)$$

$$g_{tt}(\mu, 1) = -1 + \frac{\mu}{2}.$$

Metric in the plane $y = 0$ outside the disc

In this region ($y = 0, x \geq 1$), it is also convenient to use $\tilde{\mu} = \mu(1 - x^2)$ as an abbreviation (here: $0 \geq \tilde{\mu} > -\infty$). It is easy to see that in this range $v(\mu, \tilde{\mu}) = 0$. For u and w one finds

$$u(\mu, \tilde{\mu}) = -\frac{1}{\pi} \int_0^\mu \frac{\ln(\sqrt{1 + t^2} + t)}{\sqrt{1 + t^2}} \frac{dt}{\sqrt{\mu - t}\sqrt{t - \tilde{\mu}}},$$

$$\mu w(\mu, \tilde{\mu}) = \frac{1}{\pi} \int_0^\mu \frac{\ln(\sqrt{1 + t^2} + t)}{\sqrt{1 + t^2}} \frac{\sqrt{\mu - t}}{\sqrt{t - \tilde{\mu}}} dt. \qquad (2.213)$$

Since we have not been able to solve these integrals, they remain in the following formulae

$$e^{2U} = -e^{-G} \frac{\Theta_1(\hat{u} + F(\psi, l'), l)}{\Theta_1(\hat{u} - F(\psi, l'), l)},$$

$$g_{\varrho\varrho} = \frac{2h'K(h)}{\pi} \frac{e^{2k_0} e^X e^G}{\Theta_4^2(\hat{I}(\mu), h)} \frac{\Theta_1^2(\hat{u} - F(\psi, l'), l)}{\Theta_1^2(F(\psi, l'), l)},$$

$$\Omega^2 g_{\varphi\varphi} = -(\Omega\varrho_0)^2 x^2 \hat{Q}(\hat{Q}e^{2U} - 2) + 2\Omega\varrho_0 x(\hat{Q}e^{2U} - 1) - e^{2U},$$

$$\Omega g_{\varphi t} = -\Omega\varrho_0 x(\hat{Q}e^{2U} - 1) + e^{2U},$$

$$g_{tt} = -e^{2U} \qquad (2.214)$$

with

$$\hat{u} = \sqrt[4]{(1+\mu^2)(1+\tilde{\mu}^2)}\,u,$$

$$G = \left[\frac{\pi}{K(l)}\Lambda_0(\psi,l) - \frac{\sqrt{1+\tilde{\mu}^2}+\tilde{\mu}-\mu}{\sqrt[4]{(1+\mu^2)(1+\tilde{\mu}^2)}}\right]\hat{u} + \mu w,$$

$$X = \left[\frac{\pi \hat{I}(\mu)}{2K(h)}\right]^2 \frac{\Theta_4''(0,h)}{\Theta_4(0,h)} - \left[\frac{\pi}{2K(l)}\frac{\hat{u}}{\sqrt{h'}}\right]^2 \left[\tilde{h}'l'\frac{\Theta_4''(0,l)}{\Theta_4(0,l)} - \tilde{h}l\frac{\Theta_2''(0,l)}{\Theta_2(0,l)}\right],$$

$$\hat{Q} = -\frac{4h'^2}{\pi^2}\frac{K^2(l)e^G}{\Theta_1^4(F(\psi,l'),l)}$$

$$\times \frac{\sqrt{\tilde{h}'}\Theta_4(\hat{u}-2F(\psi,l'),l) - \sqrt{\tilde{h}}\Theta_2(\hat{u}-2F(\psi,l'),l)}{\sqrt{\tilde{h}'}\Theta_4(\hat{u},l) - \sqrt{\tilde{h}}\Theta_2(\hat{u},l)}. \quad (2.215)$$

For the other quantities appearing above, see the remark after Equations (2.210); ψ is given by (2.192). Of course, in the special case $x = 1$, the above formulae for the metric potentials coincide with (2.210) for $x = 1$, in particular (2.212) holds.

Metric on the axis

For $x \to 0$, all B-periods B_{ik} diverge, but two of our combinations, namely B and R, remain finite:

$$B = B_{11} + B_{22} - 2\Re B_{12} \to \text{finite value,}$$
$$R = B_{11} - B_{22} \to \text{finite value,} \quad (2.216)$$
$$T = B_{11} + B_{22} + 2\Re B_{12} \to -\infty.$$

On the axis, we have to deal with both the case $0 < y < \Re X_2$ and the case $y > \Re X_2$. In the first case, (2.184) tells us that we have to calculate

$$f = L\frac{\vartheta_{4,3}(S-C,A-D;T,B,R) + i\vartheta_{2,1}(S-C,A-D;T,B,R)}{\vartheta_{4,3}(S+C,A+D;T,B,R) - i\vartheta_{2,1}(S+C,A+D;T,B,R)}. \quad (2.217)$$

Rosenhain's theta functions in the above equation are given by

$$\vartheta_{4,3}(S \mp C, A \mp D; T, B, R) =$$

$$\sum_{m=-\infty}^{\infty} (-1)^m \exp\left\{m^2 T + 2m(S \mp C)\right\} \vartheta_3(A \mp D + mR; B),$$

$$\vartheta_{2,1}(S \mp C, A \mp D; T, B, R) = \quad (2.218)$$

$$\sum_{m=-\infty}^{\infty} \exp\left\{\left[\tfrac{1}{2}(2m+1)\right]^2 T + (2m+1)(S \mp C)\right\}$$

$$\times \vartheta_1\left(A \mp D + \tfrac{1}{2}(2m+1)R; B\right).$$

The limiting procedure $x \to 0$ in the first equation leads to

$$\lim_{x \to 0} \vartheta_{4,3}(S \mp C, A \mp D; T, B, R) = \vartheta_3(A \mp D; B), \quad (2.219)$$

whereas in the second equation we have

$$\lim_{x \to 0} \vartheta_{2,1}(S \mp C, A \mp D; T, B, R) = \exp\left\{\tfrac{T}{4} + (C \mp S)\right\}$$

$$\times \vartheta_1(A \mp D \mp \tfrac{R}{2}; B). \quad (2.220)$$

(Note that $\lim_{x \to 0} \left(\tfrac{T}{4} + C\right)$ remains finite.) So in this case

$$f = L \frac{\vartheta_3(A - D; B) + i \exp\left\{\tfrac{T}{4} - (S - C)\right\} \vartheta_1(A - (D + \tfrac{R}{2}); B)}{\vartheta_3(A + D; B) - i \exp\left\{\tfrac{T}{4} + (S + C)\right\} \vartheta_1(A + (D + \tfrac{R}{2}); B)}, \quad (2.221)$$

and similarly in the case $y > \Re X_2$

$$f = L \frac{\vartheta_4(A - D; B) - i \exp\left\{\tfrac{T}{4} - (S - C)\right\} \vartheta_2(A - (D + \tfrac{R}{2}); B)}{\vartheta_4(A + D; B) - i \exp\left\{\tfrac{T}{4} + (S + C)\right\} \vartheta_2(A + (D + \tfrac{R}{2}); B)}. \quad (2.222)$$

On the axis, we write $u_A(\mu, y) := u(\mu, x = 0, y)$ and analogous expressions for v and w. It turns out that

$$u_A(\mu, y) = -\frac{\sqrt{\mu} y}{\pi} \int_0^\mu \frac{\ln\left(x + \sqrt{1 + x^2}\right) dx}{\sqrt{1 + x^2} \sqrt{\mu - x}(\mu[1 + y^2] - x)},$$

$$v_A(\mu, y) = y u_A(\mu, y) + \frac{I(\mu)}{\sqrt{\mu}}, \quad (2.223)$$

$$w_A(\mu, y) = y v_A(\mu, y) = y \left(y u_A(\mu, y) + \frac{I(\mu)}{\sqrt{\mu}}\right).$$

2.3 The rigidly rotating disc of dust

Therefore the functions v_A and w_A, which enter some of the arguments of formulae (2.221) and (2.222) for the Ernst potential, can be expressed by the function u_A. In addition to u_A, one other integral that could not be given in terms of elliptic functions is $\hat{I}$ (2.197), which already entered the formulae for the plane of symmetry. It is convenient to introduce the new parameter

$$\tau := \sqrt[4]{1 + \frac{1}{\mu^2}}. \tag{2.224}$$

For numerical calculations of the aforementioned remaining integrals the representation

$$\hat{I}(\mu) = \frac{2\mu\tau}{\pi} \int_0^{\frac{\pi}{2}} g(\mu \sin^2 \varphi) \sin \varphi \, d\varphi,$$

$$u_A(\mu, y) = -\frac{2}{\pi} \int_0^{\operatorname{arcsinh}(\frac{1}{y})} \frac{g\left(\mu \left[1 - y^2 \sinh^2 \alpha\right]\right)}{\cosh \alpha} d\alpha, \tag{2.225}$$

[here g is given by $g(x) = \ln(x + \sqrt{1+x^2})/\sqrt{1+x^2}$] is useful. If one calculates all the arguments of (2.221) and (2.222), splits these equations into real and imaginary parts, then one finds the following representation of the functions e^{2U} and b:

$$e^{2U}(\mu, y) = N \frac{(\tau^2 - 1)\operatorname{cn}_+ \operatorname{cn}_- + \left(\sqrt{y^4 + 2y^2 + \tau^4} - (1 + y^2)\right)}{(\tau^2 - 1)\operatorname{cn}_-^2 + \left(\sqrt{y^4 + 2y^2 + \tau^4} - (1 + y^2)\right) N^2},$$

$$b(\mu, y) = \sqrt{(\tau^2 - 1)\left(\sqrt{y^4 + 2y^2 + \tau^4} - (1 + y^2)\right)} \tag{2.226}$$

$$\times \frac{N^2 \operatorname{cn}_+ - \operatorname{cn}_-}{(\tau^2 - 1)\operatorname{cn}_-^2 + \left(\sqrt{y^4 + 2y^2 + \tau^4} - (1 + y^2)\right) N^2}.$$

Here we have made use of the notation

$$\operatorname{cn}_\pm := \operatorname{cn}\left[F(\varphi_1, h') \pm \hat{I}, h'\right], \quad h'^2 = \frac{1}{2}\left(1 - \frac{1}{\tau^2}\right) = \frac{1}{2}\left(1 - \frac{\mu}{\sqrt{1+\mu^2}}\right),$$

$$\varphi_1 := \arcsin \frac{\sqrt{2}\tau}{\sqrt{y^2 + \tau^2 + \sqrt{y^4 + 2y^2 + \tau^4}}}, \tag{2.227}$$

and the function N is defined by[13]

$$N := \exp\left\{-\mu\sqrt{y^4+2y^2+\tau^4}u_A(\mu,y)\right.$$
$$\left. -2\hat{I}(\mu)\left[Z(Y,h') + \frac{\pi Y}{2K(h)K(h')} - \frac{h'\mathrm{sn}(Y,h')\mathrm{cn}(Y,h')\mathrm{dn}(Y,h')}{1-h'\mathrm{sn}^2(Y,h')}\right.\right.$$
$$\left.\left. + \frac{\pi}{4K(h)} + \frac{y^3+\tau h'y^2+y-\tau^3 h'}{2\tau\sqrt{y^4+2y^2+\tau^4}}\right]\right\}$$

$$\times \frac{\vartheta_3\left(\frac{\pi}{2K(h)}\left(\hat{I}(\mu)+\left[Y+\frac{K(h')}{2}\right]\right),-\pi\frac{K(h')}{K(h)}\right)}{\vartheta_3\left(\frac{\pi}{2K(h)}\left(\hat{I}(\mu)-\left[Y+\frac{K(h')}{2}\right]\right),-\pi\frac{K(h')}{K(h)}\right)}, \quad (2.228)$$

with

$$Y := \mathrm{sign}(y-\tau)F(\varphi_2,h') \quad (2.229)$$

and

$$\varphi_2 := \arcsin\left[\frac{1}{h'}\sqrt{1-h\frac{\sqrt{y^4+2y^2+\tau^4}+2h\tau y}{(y+\tau)^2}}\right]. \quad (2.230)$$

Note that the moduli h and h' are the same as in the equations for the Ernst potential at the origin (2.201), (2.202), and thus depend only on μ and not on y. For the other metric functions on the axis we have

$$a = 0, \qquad e^{2k} = 1 \quad (2.231)$$

and hence the complete metric is given by

$$g_{\varrho\varrho} = e^{-2U}, \quad g_{\varphi\varphi} = 0, \quad g_{\varphi t} = 0, \quad g_{tt} = -e^{2U}. \quad (2.232)$$

A few remarks are in order:

(i) At the origin, $y = 0$, Equations (2.226)–(2.228) lead to [see (2.201) and (2.202)]

$$e^{2U}(\mu, y=0) = e^{2V_0}(\mu),$$
$$b(\mu, y=0^+) = b_0(\mu). \quad (2.233)$$

(ii) $\lim_{y\to\infty} e^{2U}(\mu,y) = 1$, $\lim_{y\to\infty} b(\mu,y) = 0$.
(iii) A series expansion of the above formulae at infinity can be used to calculate all the multipole moments of the rigidly rotating disc of dust. This will be shown in the next subsection.

---

[13] Here we use the following definition: $\mathrm{sign}(x) = 1$ for $x \geq 0$ and $\mathrm{sign}(x) = -1$ for $x < 0$.

2.3.4 Physical properties of the disc

In this subsection, we shall discuss physical quantities of the rigidly rotating disc of dust such as surface mass-density, baryonic and gravitational mass, angular momentum and also higher multipole moments. Furthermore, characteristic effects in connection with ergospheres and geodesic motion are considered.

Gravitational mass M, angular momentum J and baryonic mass M_0 can be found by specializing the corresponding formulae of Subsection 1.6.1 and calculating the resulting integrals. The fundamental quantity in this context is the surface mass-density, the derivation of which will be presented in detail using the results of Subsection 2.3.3. Alternatively, M and J can be obtained from the far field behaviour of the metric. After deriving these and the other multipoles, we shall verify that the results for M and J obtained by these different methods agree.

Later in this subsection the appearance of an ergosphere for sufficiently large μ and other characteristic relativistic effects are studied. Finally we shall discuss the Newtonian limit and the motion of test particles.

In our calculations, the parameter functions $e^{2V_0}(\mu)$, $b_0(\mu)$ and $\Omega\rho_0(\mu)$, as given in (2.201), (2.202) and (2.204), play an important role and the reader is referred back to Fig. 2.9.

Surface mass-density, mass and angular momentum

The invariant (proper) surface mass-density σ_p for the rigidly rotating disc of dust was already introduced in Subsection 1.7.3. Here we rewrite the important equations (1.103) and (1.104). The volume energy-density is given by

$$\epsilon = \sigma_p(\rho) e^{U-k} \delta(\zeta), \tag{2.234}$$

where $\delta(\zeta)$ is the usual Dirac delta distribution and σ_p can be calculated from

$$\sigma_p = \frac{1}{2\pi} e^{U-k} \frac{\partial V}{\partial \zeta}\bigg|_{\zeta=0^+}. \tag{2.235}$$

From σ_p we can calculate the total baryonic mass M_0, the gravitational mass M and the total angular momentum J using the general formulae already given in Subsection 1.6.1 together with (1.56):

$$M_0 = -\int_\Sigma \mu_B u_i n^i d\mathcal{V} = -\int_\Sigma \epsilon u_i n^i d\mathcal{V} = 2\pi e^{-V_0} \int_0^{\varrho_0} \sigma_p e^{k-U} \varrho d\varrho,$$

$$M = 2 \int_\Sigma \left(T_{ik} - \tfrac{1}{2} T^j_j g_{ik}\right) n^i \xi^k d\mathcal{V} \tag{2.236}$$

$$= 2\pi \int_0^{\varrho_0} \sigma_p e^{k-U} \varrho d\varrho + 4\pi \Omega e^{-V_0} \int_0^{\varrho_0} \sigma_p e^{k-U} u^i \eta_i \varrho d\varrho,$$

$$J = -\int_\Sigma T_{ik} n^i \eta^k d\mathcal{V} = 2\pi e^{-V_0} \int_0^{\varrho_0} \sigma_p e^{k-U} u^i \eta_i \varrho d\varrho.$$

The relation
$$M = e^{V_0} M_0 + 2\Omega J \tag{2.237}$$
follows immediately, cf. (1.157). As in Subsection 2.3.3, we use normalized coordinates $x = \varrho/\varrho_0, y = \zeta/\varrho_0$ and the combination $\tilde{\mu} = \mu(1 - x^2)$. The aim now is to express the above quantities entirely in terms of the parameter functions $e^{V_0}(\mu), b_0(\mu)$ and $\Omega \varrho_0(\mu)$ derived in Subsection 2.3.3. We first consider the two parts of the integrands, namely $\sigma \equiv \sigma_p e^{k-U}$ and $u^i \eta_i$ of Equations (2.236). The transition formulae (1.15a), (1.15b) to the corotating coordinate system[14] read

$$e^{2V} = e^{2U} \left[(1 + \Omega a)^2 - \Omega^2 \varrho^2 e^{-4U} \right],$$
$$(1 - \Omega a') e^{2V} = (1 + \Omega a) e^{2U}. \tag{2.238}$$

Differentiating the second equation with respect to y and using this relation again leads to[15]

$$V_{,y} = U_{,y} + \frac{\Omega a_{,y}}{2(1 + \Omega a)}. \tag{2.239}$$

Combining Equations (2.238), we have

$$e^{2V} e^{2U} = e^{4V} (1 - \Omega a')^2 - \Omega^2 \varrho_0^2 x^2. \tag{2.240}$$

Differentiating with respect to y and using $a'_{,y} = 0$ results in

$$2U_{,y} e^{2U} e^{2V} + 2V_{,y} e^{2U} e^{2V} = 4V_{,y} (1 + \Omega a)^2 e^{4U}. \tag{2.241}$$

Combining this with (2.239) gives

$$2\left[e^{2V} - e^{2U} (1 + \Omega a)^2 \right] V_{,y} = e^{2V} \frac{\Omega a_{,y}}{2(1 + \Omega a)}. \tag{2.242}$$

Using (1.41) in the form $\Omega a_{,y} = -\Omega \varrho_0 x e^{-4U} b_{,x}$ and the first line of (2.238) we have

$$V_{,y} = \frac{e^{2V}}{4\Omega \varrho_0 x} \frac{b_{,x}}{(1 + \Omega a) e^{2U}}. \tag{2.243}$$

---
[14] Note that $e^{2V} \equiv e^{2U'}$ and $W \equiv \varrho$.
[15] Note that $a'_{,y} = 0$, see (1.97).

2.3 The rigidly rotating disc of dust

Finally we make use of the relations $(1+\Omega a)e^{2U} = e^{V_0}(\mu)e^{V_0}(\tilde{\mu})$ and $e^{2V} = e^{2V_0}$, which are valid only within the disc, to arrive at

$$V_{,y} = \frac{b_{,x}}{2\sqrt{2\mu}xe^{V_0}(\tilde{\mu})}. \tag{2.244}$$

From (2.203) one immediately concludes that $b_{,x} = -2\mu x b'_0(\tilde{\mu})$ and hence

$$V_{,y} = -\frac{1}{2}\sqrt{2\mu}\frac{b'_0(\tilde{\mu})}{e^{V_0}(\tilde{\mu})}, \tag{2.245}$$

where a prime denotes the derivative with respect to the argument. This derivative can be calculated using (A2.27) and (A2.28). Now the surface mass-density $\sigma \equiv \sigma_p e^{k-U}$ can be written as

$$\frac{\sigma}{\Omega}(x) = \frac{1}{2\pi}\frac{1}{\Omega\varrho_0}V_{,y}(x) = -\frac{1}{2\pi}e^{-V_0}(\mu)\frac{b'_0(\tilde{\mu})}{e^{V_0}(\tilde{\mu})}. \tag{2.246}$$

The proper surface mass-density σ_p/Ω is plotted in Fig. 2.10 for several values of μ.

The remaining term is easily seen to be

$$\Omega u^i \eta_i = -e^{-V}\left[e^{2V} - e^{2U}(1+\Omega a)\right]$$
$$= -e^{-V_0}(\mu)\left[e^{2V_0}(\mu) - e^{V_0}(\mu)e^{V_0}(\tilde{\mu})\right] = e^{V_0}(\tilde{\mu}) - e^{V_0}(\mu). \tag{2.247}$$

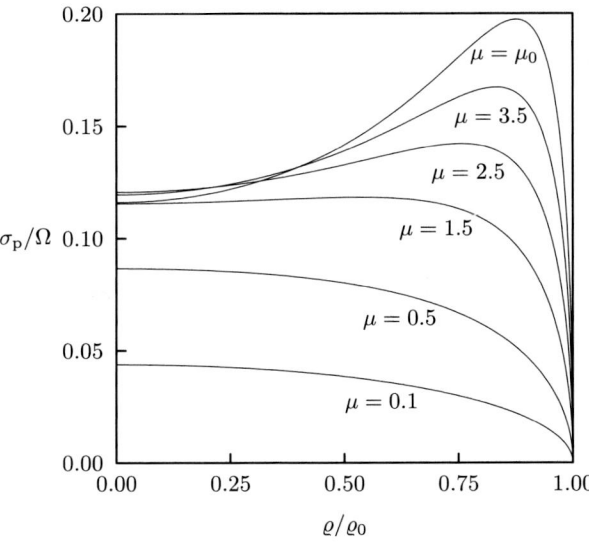

Fig. 2.10. The normalized proper surface mass-density σ_p/Ω is shown as a function of ϱ/ϱ_0 for different values of μ (after Neugebauer and Meinel 1994).

The second equality is only valid within the disc and (2.207) was used. So recalling $\mu = 2(\Omega\varrho_0)^2 e^{-2V_0}$ and $\tilde{\mu} = \mu(1-x^2)$, the baryonic mass is found to be

$$\Omega M_0(\mu) = 2\pi(\Omega\varrho_0)^2 e^{-V_0}(\mu) \int_0^1 \frac{\sigma}{\Omega} x \, dx$$

$$= -(\Omega\varrho_0)^2 e^{-2V_0}(\mu) \int_0^1 \frac{b_0'(\tilde{\mu})}{e^{V_0}(\tilde{\mu})} x \, dx \qquad (2.248)$$

$$= -\frac{1}{4} \int_0^\mu \frac{b_0'(\tilde{\mu})}{e^{V_0}(\tilde{\mu})} d\tilde{\mu}.$$

This remaining integral will be solved in the context of the multipole moments, see (2.277), (2.274) and the subsequent discussion. For the angular momentum one finds

$$\Omega^2 J(\mu) = \frac{1}{4} \int_0^\mu \left[e^{V_0}(\mu) \frac{b_0'(\tilde{\mu})}{e^{V_0}(\tilde{\mu})} - b_0'(\tilde{\mu}) \right] d\tilde{\mu}$$

$$= -e^{V_0}(\mu) \Omega M_0(\mu) - \frac{1}{4} b_0(\mu). \qquad (2.249)$$

Finally (2.237) and (2.249) lead directly to the gravitational mass

$$\Omega M(\mu) = -e^{V_0}(\mu) \Omega M_0(\mu) - \frac{1}{2} b_0(\mu). \qquad (2.250)$$

Multipole moments

The gravitational multipole moments, introduced in Geroch (1970), were generalized to the case of asymptotically flat stationary space-times in Hansen (1974) and Thorne (1980). In the case of axisymmetric systems, Fodor *et al.* (1989) developed a scheme for calculating the multipole moments by expanding the Ernst potential along the axis of symmetry. Using this scheme, the moments of the rigidly rotating disc of dust were calculated by Kleinwächter *et al.* (1995). For calculating the expansion coefficients of the Ernst potential, it was, however, necessary to integrate over functions, which had to be determined as solutions of an integral equation. Here we shall provide an explicit scheme, again following the algorithm of Fodor *et al.* (1989), for calculating all the multipole moments as given functions of the parameter μ.

The starting point for the calculation is a power series expansion of the function

$$g(\varrho = 0, \zeta) := \zeta \frac{1 - f(0, \zeta)}{1 + f(0, \zeta)} = \sum_{n=0}^{\infty} \frac{m_n}{\zeta^n}, \qquad \zeta > 0 \qquad (2.251)$$

at infinity, where $f(0, \zeta)$ is the Ernst potential on the axis. The multipole moments

$$P_n = M_n + i J_n \qquad (2.252)$$

2.3 The rigidly rotating disc of dust

can be obtained as functions of the coefficients m_n of this expansion. The real parts of the P_n are called the mass moments M_n and the imaginary parts the rotational moments J_n. It turns out that, as in the case of the m_n, all P_n with even index are real and the others are purely imaginary. This is a general property of solutions with reflectional symmetry (Kordas 1995, Meinel and Neugebauer 1995). Thus we have

$$M_l = 0, \ l = 1, 3, 5, \ldots \quad \text{and} \quad J_l = 0, \ l = 0, 2, 4, \ldots. \quad (2.253)$$

Here $M_n|_{n=0} = M$ is the gravitational mass (not to be confused with the baryonic mass) and $J_1 = J$ is the ζ-component of the angular momentum.

Inserting the Ernst potential $f = e^{2U} + ib$ into (2.251) yields

$$\begin{aligned} g(x = 0, y) &= \varrho_0 \, y \frac{1 - f(0, y)}{1 + f(0, y)} \\ &= \varrho_0 \, y \frac{1 - (e^{4U} + b^2)}{(1 + e^{2U})^2 + b^2} - i \frac{2\varrho_0 \, yb}{(1 + e^{2U})^2 + b^2}. \end{aligned} \quad (2.254)$$

Using the dimensionless coefficients $\tilde{m}_n := \Omega^{n+1} m_n$, the power series expansion (2.251) reads

$$g(0, y) = \sum_{n=0}^{\infty} \varrho_0 \frac{\tilde{m}_n}{(\Omega\varrho_0)^{n+1} y^n}. \quad (2.255)$$

For the Ernst potential on the axis, a slightly different representation from (2.226) will be inserted in (2.254), where again the parameter functions $b_0(\mu)$ and $\Omega\varrho_0(\mu)$ will be used:[16]

$$\begin{aligned} f(x = 0, y) &= \frac{N(1 - \mathcal{MP})}{N^2 + \mathcal{P}^2} - i \frac{\mathcal{M}N^2 + \mathcal{P}}{N^2 + \mathcal{P}^2}, \\ \mathcal{M} &= \frac{1 - 2(1 + y^2 + \sqrt{y^4 + 2y^2 + \tau^4})(\Omega\varrho_0)^2}{2(\Omega\varrho_0)y - b_0}, \\ \mathcal{P} &= -\frac{1 - 2(1 + y^2 + \sqrt{y^4 + 2y^2 + \tau^4})(\Omega\varrho_0)^2}{2(\Omega\varrho_0)y + b_0}. \end{aligned} \quad (2.256)$$

The function $N = N(\mu, y)$ is given in (2.228) and can be expanded as follows:

$$N = \exp\left(\frac{c_1}{y} + \frac{c_3}{y^3} + \frac{c_5}{y^5} + \frac{c_7}{y^7} + \ldots\right). \quad (2.257)$$

---

[16] This form of the axis potential is discussed by Kleinwächter (2000). Because b_0 is negative, there is a zero in the denominator of $\mathcal{P}$, which is compensated for by the numerator, however.

Carrying out the expansion of (2.254) we can read off the coefficients $\tilde{m}_n(\mu)$ as functions of the parameter b_0, the newly defined

$$\Omega_0 := \Omega \varrho_0 \tag{2.258}$$

and the above coefficients c_n, which all depend only on μ. The expansion of (2.254) leads to

$$g = \varrho_0 \left(\tilde{m}_0 + \frac{\tilde{m}_1}{y} + \frac{\tilde{m}_2}{y^2} + \frac{\tilde{m}_3}{y^3} + \frac{\tilde{m}_4}{y^4} + \cdots \right),$$

$$\tilde{m}_0 = -b_0 - \Omega_0 c_1,$$

$$\tilde{m}_1 = -i(b_0 + 2\Omega_0 c_1),$$

$$\tilde{m}_2 = b_0 + (2 + b_0^2)\Omega_0 c_1 + b_0 \Omega_0^2 c_1^2 + (\Omega_0^3(c_1^3 - 12c_3))/3,$$

$$\tilde{m}_3 = (i/3)(3b_0^3 + 12b_0^2 \Omega_0 c_1 + 3b_0 \Omega_0^2(4 + 3c_1^2)$$
$$+ 2\Omega_0^3(12c_1 + c_1^3 - 12c_3)),$$

$$\tilde{m}_4 = b_0 - 2b_0^3 + (2 - 7b_0^2)\Omega_0 c_1 - b_0 \Omega_0^2(8 + (5 + b_0^2)c_1^2)$$
$$- (2\Omega_0^3(24c_1 + (1 + 2b_0^2)c_1^3 - 6(2 + b_0^2)c_3))/3$$
$$+ b_0 \Omega_0^4((-2c_1^4)/3 + 8c_1 c_3) + \Omega_0^5((-2c_1^5)/15 + 4c_1^2 c_3 - 16c_5). \tag{2.259}$$

It is possible to calculate the coefficients $c_n(\mu)$ up to arbitrary order and hence the same is valid for the quantities $\tilde{m}_n$. Instead of the multipole moments (2.252), we use the dimensionless and normalized moments

$$\tilde{P}_n := \tilde{M}_n + i\tilde{J}_n; \quad n = 0, 1, 2, \ldots \quad \text{with:}$$
$$\tilde{M}_{2l} := (-1)^l (2\Omega)^{2l+1} M_{2l}, \tag{2.260}$$
$$\tilde{J}_{2l+1} := (-1)^l (2\Omega)^{2l+2} J_{2l+1}; \quad l = 0, 1, 2, \ldots.$$

It turns out that

$$\tilde{P}_0 = \tilde{m}_0, \quad \tilde{P}_1 = \tilde{m}_1, \quad \tilde{P}_2 = -\tilde{m}_2, \quad \tilde{P}_3 = -\tilde{m}_3 \tag{2.261}$$

holds, thus implying

$$\tilde{M}_0 = \tilde{m}_0, \quad \tilde{J}_1 = \Im \tilde{m}_1, \quad \tilde{M}_2 = -\tilde{m}_2, \quad \tilde{J}_3 = -\Im \tilde{m}_3 \tag{2.262}$$

(i.e. $|\tilde{P}_n| = |\tilde{m}_n|$, $n = 0, 1, 2, 3$). But for $i > 3$ one has the general structure

$$|\tilde{P}_n| = |\tilde{m}_n + f_n(\tilde{m}_j : j < n)|. \tag{2.263}$$

The exact form of the functions f_n up to $n = 10$ is given in Fodor *et al.* (1989) and can be calculated with the methods given there up to arbitrary order.[17] Summarizing the above we have

$$\tilde{P}_n(\mu) = \tilde{P}_n\left[b_0(\mu), \Omega_0(\mu), c_j(\mu) : j < j_{\max}(n)\right], \quad (2.264)$$

where $j_{\max}(n)$ refers to the finite maximal value of j for a given n. Because the coefficients c_j can also be obtained as functions of μ for arbitrary j, all multipole moments of the rigidly rotating disc of dust can be calculated explicitly. From the structure of the function g, it follows that all the coefficients $\tilde{m}_n$ with an even index are real and all the others are purely imaginary. This property transfers to the $\tilde{P}_n$. For all $\mu < \mu_0$ and $n > 3$, the quantities $|\tilde{P}_n|$ differ from the absolute values $|\tilde{m}_n|$. But in the limit $\mu \to \mu_0$, for which $\Omega_0 \to 0$ and $b_0 \to -1$ hold, $|\tilde{P}_i(\mu_0)| = |\tilde{m}_i(\mu_0)| = 1 (i = 0, 1, 2, \ldots)$, or more precisely

$$\tilde{M}_{2l}(\mu_0) = (-1)^l \tilde{m}_{2l}(\mu_0) \quad = 1, \quad (2.265)$$

$$\tilde{J}_{2l+1}(\mu_0) = (-1)^l \Im \tilde{m}_{2l+1}(\mu_0) = 1 \quad (2.266)$$

is valid for all l. These are exactly the multipole moments of the extreme Kerr solution (in our normalization).[18] We shall come back to this important point in Subsection 2.3.5.

The remaining task is to calculate the coefficients c_n of (2.257). The function $\ln N$ is built up from the functions u_A (2.223) and the function Y, which appears as the argument of different elliptic functions (2.228). The expansion of the term involving u_A can be expressed with the integrals[19]

$$I_k(\mu) := \frac{1}{\pi} \int_0^\mu \frac{\ln(\sqrt{1+x^2}+x)}{\sqrt{1+x^2}} \frac{x^k}{\sqrt{\mu - x}} dx. \quad (2.267)$$

As an illustration, we provide the first terms of the expansion

$$-\mu\sqrt{y^4 + 2y^2 + \tau^4} u_A(\mu, y) = \sqrt{\mu} I_0(\mu) y + \frac{1}{\sqrt{\mu}} I_1(\mu) \frac{1}{y} + \frac{I_0 - 2\mu I_1 + 2I_2}{2\mu^{3/2} y^3} + \mathcal{O}\left(\frac{1}{y^5}\right). \quad (2.268)$$

For the function Y defined in (2.229),

$$Y = \frac{K(h')}{2} + \frac{1}{y} + \frac{1 - 2h^2}{3y^3} + \frac{1 - 6h^2 + 6h^4}{5y^2} + \mathcal{O}\left(\frac{1}{y^7}\right) \quad (2.269)$$

---

[17] For $n = 4, 5, 6$ the functions f_n can be found in Appendix 3.
[18] For a more detailed discussion see Appendix A3.3.
[19] Note the identity $I_0(\mu) = I(\mu)$.

follows directly. From this series, one can in turn obtain the series expansions of the Jacobian zeta function $Z(Y, h')$ and the elliptic functions $\text{sn}(Y, h'), \text{cn}(Y, h')$ and $\text{dn}(Y, h')$. Finally we calculate the logarithm of the theta quotient

$$\ln TQ := \ln \frac{\vartheta_3\left(\frac{\pi}{2\text{K}(h)}\left(\hat{I}(\mu) + \left[Y + \frac{\text{K}(h')}{2}\right]\right), -\pi\frac{\text{K}(h')}{\text{K}(h)}\right)}{\vartheta_3\left(\frac{\pi}{2\text{K}(h)}\left(\hat{I}(\mu) - \left[Y + \frac{\text{K}(h')}{2}\right]\right), -\pi\frac{\text{K}(h')}{\text{K}(h)}\right)}. \tag{2.270}$$

Introducing $\hat{Y} := Y - \text{K}(h')/2$ and using (A2.7), one can write

$$\ln TQ = \ln \vartheta_2\left(\frac{\pi}{2\text{K}(h)}\left(\hat{I}(\mu) + \hat{Y}\right), -\pi\frac{\text{K}(h')}{\text{K}(h)}\right) \\ - \ln \vartheta_2\left(\frac{\pi}{2\text{K}(h)}\left(\hat{I}(\mu) - \hat{Y}\right), -\pi\frac{\text{K}(h')}{\text{K}(h)}\right) + \frac{\pi}{\text{K}(h)}\hat{I}(\mu). \tag{2.271}$$

Using (A2.29) it is possible to perform the expansion of the last equation. The coefficients of this series consist of the complete elliptic integrals $\text{K}(h')$ and Jacobian elliptic functions with argument $\hat{I}(\mu)$ and modulus h'. Inserting all these series for the components of $\ln N$, we can derive the desired coefficients c_n. These coefficients have the structure

$$c_n = c_n\left[\mu, I_k(\mu) : k \leq k_{\max}(n)\right]. \tag{2.272}$$

In addition to a simple algebraic dependence on μ and $I_k(\mu)$ (2.267), c_n depends on elliptic functions and elliptic integrals with main argument $\hat{I}(\mu) = \sqrt[4]{1 + \mu^2}I_0(\mu)$ and modulus $h'(\mu)$. The explicit results for c_1, c_3, c_5, c_7, c_9 and c_{11}, which are needed for the first eleven multipole moments are given in Appendix 3.

The coefficient c_1 is especially interesting since it is the only one that enters the formulae for the first two moments yielding the gravitational mass and the angular momentum of the disc. This coefficient reads

$$c_1 = \frac{1}{\sqrt{\mu}}\left\{2\sqrt{1 + \mu^2}I(\mu)\left[h'^2 - \frac{\text{E}(h)}{\text{K}(h)}\right] + I_1(\mu) \\ + \sqrt[4]{1 + \mu^2}\frac{\pi}{\text{K}(h)}\Lambda_0(\text{am}(\sqrt[4]{1 + \mu^2}I(\mu), h'), h)\right\}, \tag{2.273}$$

which can be simplified using $\psi_1 := \text{am}(\hat{I}(\mu), h')$

$$c_1(\mu) = \frac{1}{\sqrt{\mu}}\left\{\frac{\sqrt{2}}{\sqrt{hh'}}\left[\text{E}(\psi_1, h') - h^2\hat{I}(\mu)\right] + I_1(\mu)\right\}. \tag{2.274}$$

2.3 The rigidly rotating disc of dust

Thus for the normalized gravitational mass ΩM and ζ-component of the angular momentum $\Omega^2 J$ we have,

$$\Omega M(\mu) = -\frac{1}{2}\left[c_1(\mu)\Omega_0(\mu) + b_0(\mu)\right], \tag{2.275}$$

$$\Omega^2 J(\mu) = -\frac{1}{2}\left[c_1(\mu)\Omega_0(\mu) + \frac{1}{2}b_0(\mu)\right]. \tag{2.276}$$

From the relation (2.237) between ΩM, $\Omega^2 J$ and ΩM_0, one immediately finds

$$\Omega M_0(\mu) = \frac{1}{4}\sqrt{2\mu}c_1(\mu). \tag{2.277}$$

On the other hand we have the result (2.248) for ΩM_0 which was obtained by integrating over the proper surface mass-density. This means that the identity

$$\frac{d}{d\mu}\left[\sqrt{2\mu}c_1(\mu)\right] = -\frac{b_0'(\mu)}{e^{V_0(\mu)}} \tag{2.278}$$

must hold, which can easily be verified by differentiating (2.274) with the help of (A2.26) to provide an expression for the left hand side. The right hand side can be found by differentiating b_0, taking into account (A2.27) and (A2.28). Conversely, we have thus solved for the integral in (2.248) and have proved that the results that were obtained in connection with the multipole expansion (2.275), (2.276) and (2.277) coincide exactly with the results (2.248), (2.249) and (2.250) calculated by integrating over the source. The relative binding energy is

$$\frac{M_0 - M}{M_0}(\mu) = 1 + e^{V_0}(\mu) + \frac{2b_0(\mu)}{\sqrt{2\mu}c_1(\mu)}. \tag{2.279}$$

For $\mu = \mu_0$, one finds $\Omega_0 = 0$, $b_0 = -1$ and $c_1 = 1.0487435\ldots$ Therefore, in the extreme relativistic limit, we have the values

$$\Omega M(\mu_0) = \frac{1}{2}, \qquad \Omega^2 J(\mu_0) = \frac{1}{4} \tag{2.280}$$

and

$$\Omega M_0(\mu_0) = 0.797809010\ldots, \qquad \frac{M_0 - M}{M_0}(\mu_0) = 0.373283588\ldots \tag{2.281}$$

The normalized mass ΩM (2.275), angular momentum $\Omega^2 J$ (2.276) and baryonic mass ΩM_0 (2.277) are plotted as functions of μ in Fig. 2.11, and the graphs of the relative binding energy (2.279) and of the ratio M^2/J can be found in Fig. 2.12.

The first six mass multipole moments are plotted in Fig. 2.13 and the first five rotational moments are shown in Fig. 2.14. The complete corresponding formulae are given in Appendix 3.

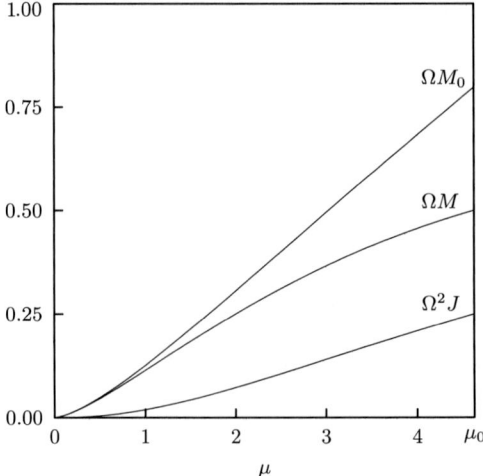

Fig. 2.11. The normalized baryonic mass ΩM_0, gravitational mass ΩM and angular momentum $\Omega^2 J$ are given as functions of μ (after Neugebauer *et al.* 1996).

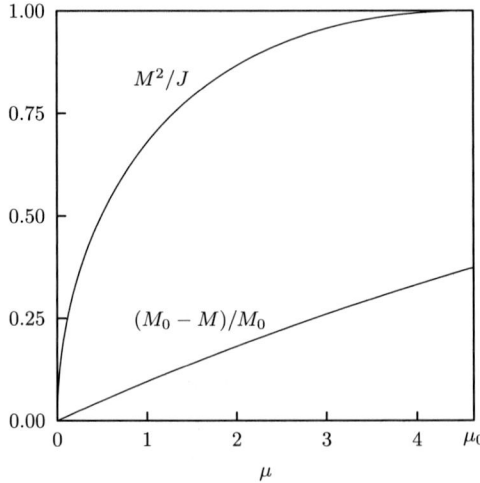

Fig. 2.12. Relative binding energy and M^2/J (after Neugebauer *et al.* 1996).

The ergosphere

For sufficiently large values of μ, the rigidly rotating disc of dust possesses an ergosphere, i.e. a region in which no static observer (seen from infinity) can reside, see Subsection 1.6.2. Within this region $d\varphi/dt > 0$ must hold for any timelike worldline. The ergosphere is characterized by

$$\xi_i \xi^i = g_{44} > 0, \tag{2.282}$$

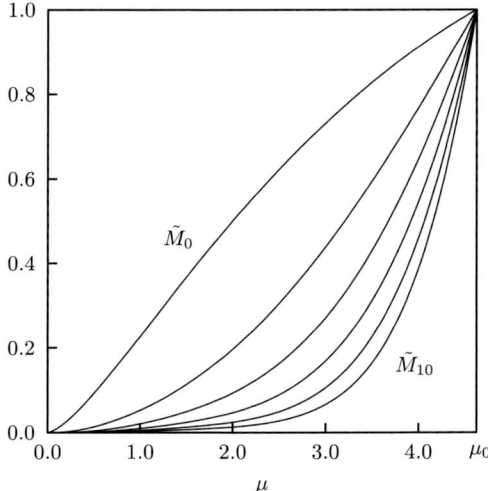

Fig. 2.13. The normalized mass multipole moments $\tilde{M}_0, \tilde{M}_2, \tilde{M}_4, \tilde{M}_6, \tilde{M}_8$ and $\tilde{M}_{10}$ (after Kleinwächter *et al.* 1995).

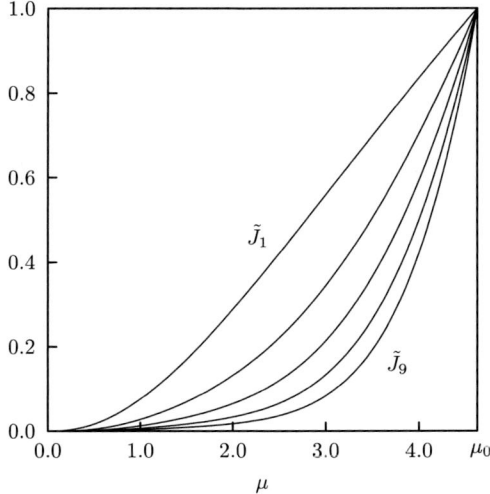

Fig. 2.14. The normalized rotational multipole moments $\tilde{J}_1, \tilde{J}_3, \tilde{J}_5, \tilde{J}_7$ and $\tilde{J}_9$ (after Kleinwächter *et al.* 1995).

meaning that the Killing vector ξ_i of stationarity [normalized by (1.3)] becomes spacelike there.

The occurrence of an ergosphere is observed for

$$\mu > \mu_e = 1.68849467\ldots \tag{2.283}$$

For $\mu = \mu_e$, we find a ring at $x_e = (\varrho/\varrho_0)_e = 0.7617018\ldots$, where $g_{44} = 0$ holds. Values for both μ_e and x_e can be obtained simultaneously by searching for the first zero $e^{2U}(\mu_e, x_e) = 0$ using (2.203).

From the last line of (2.212) it is obvious that the rim of the disc belongs to the ergosphere for $\mu > 2$. For $\mu \to \mu_0$, the shape of the ergosphere of the extreme Kerr solution is approached. The general torus-like shape can be obtained by searching for the zeros of the numerators of e^{2U} in (2.185) and (2.186). Five characteristic shapes of the ergosphere are shown in Fig. 2.15.

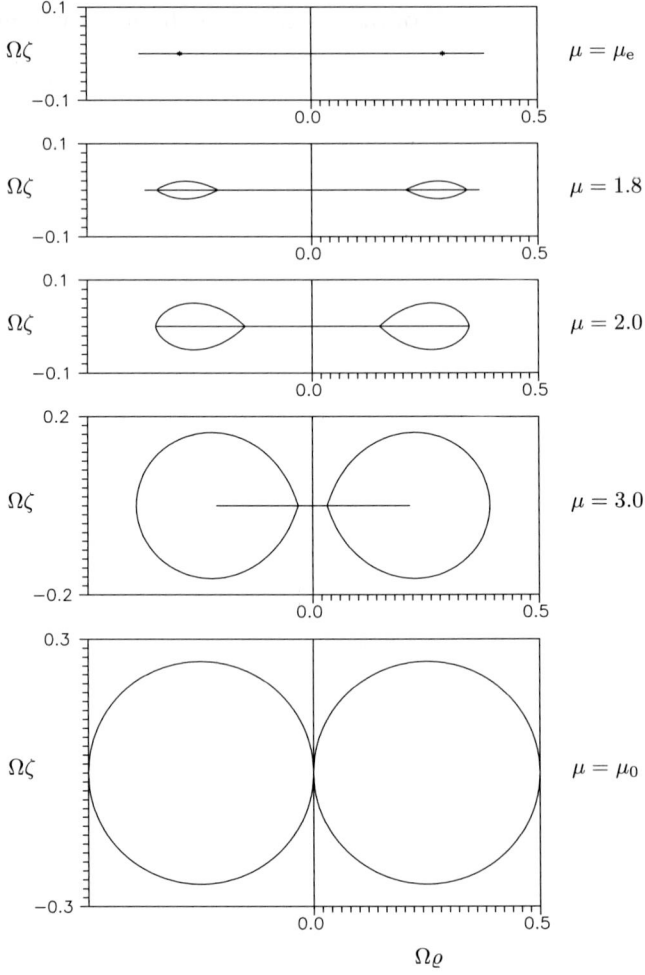

Fig. 2.15. Ergospheres for different values of μ. An ergosphere appears for $\mu > \mu_e = 1.68849\ldots$ and reaches the rim of the disc at $\mu = 2$. For $\mu \to \mu_0$ the ergosphere of the extreme Kerr solution (with $M = 1/2\Omega$) is approached. The horizontal lines represent the disc. Its coordinate radius ϱ_0 vanishes in the limit $\mu \to \mu_0$, see Subsection 2.3.5 (after Meinel and Kleinwächter 1995).

Further characteristic relativistic effects

We start by considering circular geodesic motion at the rim of the disc, which can be solved particularly elegantly. For $x = 1$, we find from (2.210) the expressions for the required metric coefficients and their derivatives:

$$\Omega^2 g_{\varphi\varphi} = -(1 - e^{V_0})^2 + \frac{\mu}{2}, \quad \Omega g_{\varphi t} = 1 - e^{V_0} - \frac{\mu}{2}, \quad g_{tt} = -1 + \frac{\mu}{2}, \quad (2.284)$$

$$\Omega^2 g_{\varphi\varphi,x} = \mu(2e^{V_0} - 1), \quad \Omega g_{\varphi t,x} = \mu(1 - e^{V_0}), \quad g_{tt,x} = -\mu, \quad (2.285)$$

where $e^{V_0}(\mu)$ is given by (2.209). From (2.284) we immediately obtain the angular velocity ω, as seen from infinity, of the so-called locally non-rotating observer[20]

$$\omega = -\frac{g_{\varphi t}}{g_{\varphi\varphi}} = \Omega \frac{\mu - 2(1 - e^{V_0})}{\mu - 2(1 - e^{V_0})^2}. \quad (2.286)$$

The value of ω approaches Ω in the extreme relativistic limit $\mu \to \mu_0$ (i.e. $e^{V_0} \to 0$). This property holds not only at the rim of the disc, but throughout the whole disc as we can see by rewriting (2.210)

$$\omega(\mu, x) = \Omega \frac{\mu x^2 + 2e^{V_0}(\tilde{\mu})\left[e^{V_0}(\mu) - e^{V_0}(\tilde{\mu})\right]}{\mu x^2 - 2\left[e^{V_0}(\mu) - e^{V_0}(\tilde{\mu})\right]^2}. \quad (2.287)$$

We have to distinguish between corotating particles (direct orbits: $\Omega_+ > 0$) and counter-rotating particles (retrograde orbits: $\Omega_- < 0$). Of course, $\Omega_+ = \Omega$, since the dust particles of the disc move on circular geodesics themselves. For Ω_- the result is

$$\Omega_- = -\frac{\Omega}{2e^{V_0} - 1}. \quad (2.288)$$

The corresponding specific energies $E_\pm = -\xi_i u_\pm^i$ and the specific angular momenta $L_\pm = \eta_i u_\pm^i$ (where $u_\pm^i$ is the corresponding 4-velocity) turn out to be

$$E_+ = 1, \quad L_+ = \frac{1}{\Omega}(1 - e^{V_0}), \quad (2.289)$$

$$E_- = \frac{1 - \mu}{\sqrt{1 - 2\mu}}, \quad L_- = -\frac{1}{\Omega}\frac{\mu - 1 + e^{V_0}}{\sqrt{1 - 2\mu}}. \quad (2.290)$$

For the linear velocity of rotation $v_\pm$ measured in the locally non-rotating frame of reference we get from (1.32)

$$v_+ = \sqrt{\frac{2}{\mu}}(1 - e^{V_0}), \quad v_- = -\sqrt{\frac{2}{\mu}}\frac{\mu - 1 + e^{V_0}}{2e^{V_0} - 1}, \quad (2.291)$$

---

[20] See Section 1.4, especially footnote 6.

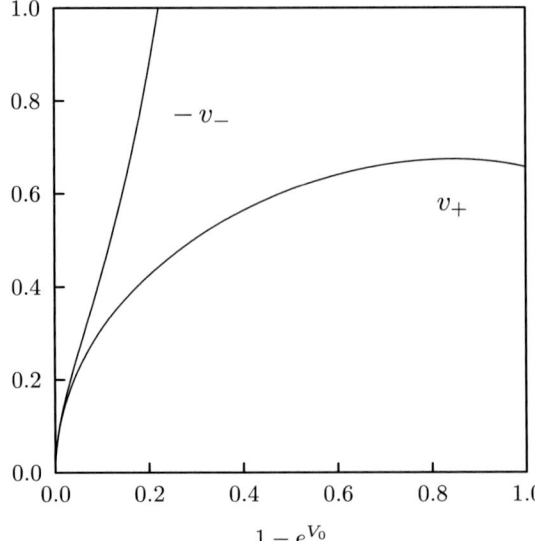

Fig. 2.16. The linear velocities of rotation $v_\pm$ of test particles moving on circular geodesic orbits at the rim of the disc measured in the locally non-rotating frame of reference. Retrograde motion (v_-) is possible only for $\mu < 1/2$ (after Meinel and Kleinwächter 1995).

a plot of which can be found in Fig. 2.16. It turns out that for $\mu \geq 1/2$ no *geodesic* retrograde motion is possible. From (2.290) it is obvious that E_- and L_- become infinite and (2.291) shows that $|v_-|$ approaches 1, i.e. the speed of light in our units, for $\mu = 1/2$. As we already pointed out, for $\mu \geq 2$ the rim of the disc belongs to the ergosphere, meaning that for such μ, no motion at all of a particle or observer in the retrograde direction is possible there.

Figure 2.17 combines the parameter relations between ΩM and M^2/J for the rigidly rotating disc of dust as well as for the Kerr black hole (1.136). The two branches meet at the point $M^2/J = 1$. In this figure, the corresponding relation for the Newtonian disc, which follows from (2.24) and (2.26), is also plotted. We see right away that in the extremely relativistic regime, the curve for the relativistic disc of dust differs from its Newtonian counterpart significantly. We observe the parametric transition to the Kerr black hole for $M^2/J = 1$ ($\mu \to \mu_0$), see Subsection 2.3.5. The Einsteinian disc possesses characteristic relativistic features even in regions where the curves for both discs differ by only very little. For $M^2/J(\mu = 1/2) = 0.50457\ldots$, where as discussed no geodesic retrograde motion at the rim is possible, both curves nearly coincide and even for $M^2/J(\mu = 2) = 0.86656\ldots$, where no retrograde motion at all is possible, one finds only a moderate deviation.

2.3 The rigidly rotating disc of dust

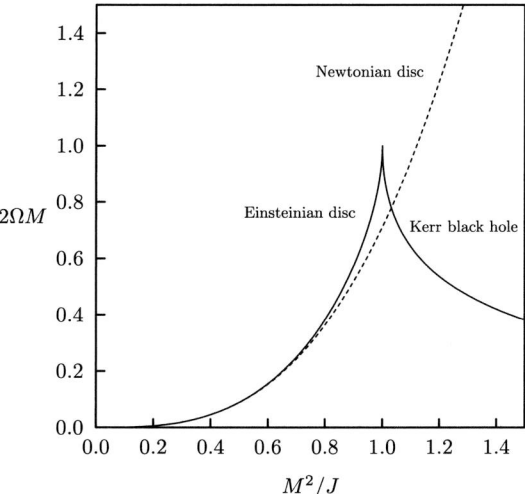

Fig. 2.17. Relation between ΩM and M^2/J for the classical Maclaurin disc (dashed line), the general relativistic disc of dust and the Kerr black hole. In the latter case Ω has been identified with the 'angular velocity' Ω_h of the horizon, see Subsection 1.8.2 (after Neugebauer and Meinel 1993).

The Newtonian limit

Of course, for all functions discussed here, such as the parameter functions $e^{2V_0}, b_0, \Omega_0 \equiv \Omega\varrho_0$ or the mass M and angular momentum J as well as the higher multipole moments, a Taylor expansion around the Newtonian limit can be given. For example, from (2.209) and (2.204) we can derive

$$e^{V_0}(\mu) = 1 - \frac{\mu}{2} + \frac{\mu^2}{8} + \left(\frac{1}{16} - \frac{8}{9\pi^2}\right)\mu^3 + \left(-\frac{5}{128} + \frac{4}{9\pi^2}\right)\mu^4 + \mathcal{O}(\mu^5),$$

$$\Omega_0(\mu) = \frac{\sqrt{\mu}}{\sqrt{2}} - \frac{\mu^{3/2}}{2\sqrt{2}} + \frac{\mu^{5/2}}{8\sqrt{2}} + \frac{(9-\frac{128}{\pi^2})\mu^{7/2}}{144\sqrt{2}} + \frac{(-45+\frac{512}{\pi^2})\mu^{9/2}}{1152\sqrt{2}} + \mathcal{O}\left(\mu^{\frac{11}{2}}\right)$$
(2.292)

and (2.275), (2.276), (2.277) lead to

$$\Omega M(\mu) = \frac{\sqrt{2}\mu^{3/2}}{3\pi} - \frac{\mu^{5/2}}{5\sqrt{2}\pi} + \frac{19\mu^{7/2}}{420\sqrt{2}\pi} + \frac{(-640+33\pi^2)\mu^{9/2}}{3240\sqrt{2}\pi^3} + \mathcal{O}\left(\mu^{\frac{11}{2}}\right),$$

$$\Omega M_0(\mu) = \frac{\sqrt{2}\mu^{3/2}}{3\pi} - \frac{\sqrt{2}\mu^{5/2}}{15\pi} + \frac{\mu^{7/2}}{35\sqrt{2}\pi} + \frac{16\sqrt{2}(-35+3\pi^2)\mu^{9/2}}{2835\pi^3} + \mathcal{O}\left(\mu^{\frac{11}{2}}\right),$$

$$\Omega^2 J(\mu) = \frac{\sqrt{2}\mu^{5/2}}{15\pi} - \frac{\mu^{7/2}}{15\sqrt{2}\pi} + \frac{(896-39\pi^2)\mu^{9/2}}{2268\sqrt{2}\pi^3} + \mathcal{O}\left(\mu^{\frac{11}{2}}\right).$$
(2.293)

In order to compare these with the corresponding Newtonian formulae it is instructive to invert the series for Ω_0:[21]

$$\mu(\Omega_0) = 2\Omega_0^2 + 4\Omega_0^4 + 12\Omega_0^6 + \left(40 + \frac{256}{9\pi^2}\right)\Omega_0^8$$
$$+ \left(140 + \frac{2560}{9\pi^2}\right)\Omega_0^{10} + \mathcal{O}(\Omega_0^{12}) \qquad (2.294)$$

and insert this result in the above series

$$\Omega M(\Omega_0) = \frac{4\Omega_0^3}{3\pi} + \frac{16\Omega_0^5}{5\pi} + \frac{1088\Omega_0^7}{105\pi} + \frac{256(40 + 57\pi^2)\Omega_0^9}{405\pi^3} + \mathcal{O}(\Omega_0^{11}),$$

$$\Omega M_0(\Omega_0) = \frac{4\Omega_0^3}{3\pi} + \frac{52\Omega_0^5}{15\pi} + \frac{1214\Omega_0^7}{105\pi} + \frac{2(31360 + 58791\pi^2)\Omega_0^9}{2835\pi^3} + \mathcal{O}(\Omega_0^{11}),$$

$$\Omega^2 J(\Omega_0) = \frac{8\Omega_0^5}{15\pi} + \frac{32\Omega_0^7}{15\pi} + \frac{128(140 + 177\pi^2)\Omega_0^9}{2835\pi^3} + \mathcal{O}(\Omega_0^{11}). \qquad (2.295)$$

Comparing with (2.24) for M and with (2.26) for J (remembering $\Omega_0 = \Omega \varrho_0$), we observe that the leading terms coincide with the expressions for the Maclaurin disc. In the same way, the proper surface-mass density

$$\sigma_p(\Omega_0, x) = \frac{2\sqrt{1-x^2}\Omega_0}{\pi^2} + \frac{4\sqrt{1-x^2}(x^2+2)\Omega_0^3}{3\pi^2}$$
$$+ \frac{4\sqrt{1-x^2}(3x^4 + 14x^2 + 28)\Omega_0^5}{15\pi^2} + \mathcal{O}(\Omega_0^7) \qquad (2.296)$$

can be seen to coincide with (2.23) to leading order by taking (2.24) and $x = \varrho/\varrho_0$ into account. In

$$V_0(\Omega_0) = -\Omega_0^2 - 2\Omega_0^4 + \frac{16}{9}\left(-3 - \frac{4}{\pi^2}\right)\Omega_0^6$$
$$+ \left(-16 - \frac{512}{9\pi^2}\right)\Omega_0^8 + \mathcal{O}(\Omega_0^{10}) \qquad (2.297)$$

---

[21] Note that Ω_0 is not a monotonic function in μ (see Fig. 2.9). Hence an inversion is not possible in the whole range ($0 \leq \mu \leq \mu_0$). Near the Newtonian limit ($\mu \ll 1$), this presents no problem, however.

the parameter relation (2.25) of the Maclaurin disc is also found to leading order. Finally, we consider the power series expansion for the relative binding energy

$$\frac{M_0 - M}{M_0}(\mu) = \frac{\mu}{10} - \frac{\mu^2}{200} + \left(\frac{3809}{126000} - \frac{8}{27\pi^2}\right)\mu^3 + \mathcal{O}(\mu^4),$$

$$\frac{M_0 - M}{M_0}(\Omega_0) = \frac{\Omega_0^2}{5} + \frac{19\Omega_0^4}{50} + \left(\frac{21449}{15750} - \frac{64}{27\pi^2}\right)\Omega_0^6 + \mathcal{O}(\Omega_0^8).$$

(2.298)

The leading term of the last series gives the Newtonian value $-(E_{\text{rot}} + E_{\text{g}})/M$ as follows from (2.27) and (2.24).

An impression of the accuracy of the above formulae can be gained from Fig. 2.18, in which the expressions (2.293) are compared with the corresponding exact curves for the range $0 \leq \mu \leq 3$.

A systematic post-Newtonian expansion of the whole solution is given in Petroff and Meinel (2001). There, as in Bardeen and Wagoner (1971), Padé approximants were used to obtain a more accurate representation of the physical parameters from their truncated series representations.

Motion of test particles

In the context of the study of physical spacetime properties, it is natural to investigate the relativistic motion of test particles in the vicinity of the gravitational source. Here we concentrate on circular orbits within the plane of the disc, centred about

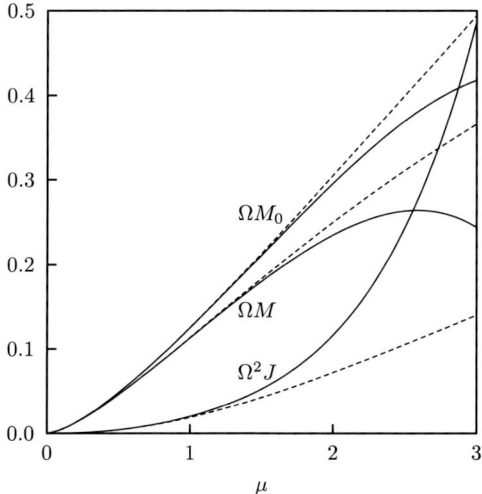

Fig. 2.18. The post-Newtonian expansions of ΩM, ΩM_0 and $\Omega^2 J$ up to the order given in (2.293), as compared to the exact curves (dashed lines), cf. Fig. 2.11.

the rotational axis.[22] One finds that there is always a (stable or unstable) circular orbit for positive angular momentum and a given radius. However, for sufficiently relativistic discs (the more precise condition follows below), there are regions within the plane of the disc in which a particle with negative angular momentum cannot follow a circular path. If the disc is still more strongly relativistic, then one finds circular orbits with negative energies of arbitrary magnitude.

We start the study of geodesic motion by considering the corresponding Hamiltonian system, which reads

$$\mathcal{H} = \frac{1}{2} g^{ij} p_i p_j \quad \text{with} \quad p_i = g_{ij} \dot{x}^j, \quad \left(\dot{} = \frac{\mathrm{d}}{\mathrm{d}\tau}, \quad \tau : \text{proper time} \right). \quad (2.299)$$

For the timelike geodesic equations

$$\ddot{x}^i + \Gamma^i_{kl} \dot{x}^k \dot{x}^l = 0, \quad \dot{x}_k \dot{x}^k = -1, \quad (2.300)$$

this system can be reduced, for axisymmetry and stationarity, to a conservative Hamiltonian of two degrees of freedom of the form

$$\mathcal{H} = \frac{1}{2} e^{2(U-k)} (p_\varrho^2 + p_\zeta^2) + \frac{1}{2} \left[1 - \frac{1}{\varrho^2} (L^2 g_{tt} + 2LE g_{\varphi t} + E^2 g_{\varphi\varphi}) \right]. \quad (2.301)$$

The constants of motion L and E have the same meaning as they did a few pages ago, i.e. $L = p_\varphi = $ constant is the (ζ-component of the) specific angular momentum of the particle and $E = -p_t = $ constant its specific energy.

The above Hamiltonian is invariant under a simultaneous change of the signs of L and E. We fix these signs by the condition $\dot{t} > 0$ and this condition is then satisfied along the whole trajectory.

The Hamiltonian (2.301) is chosen such that $\mathcal{H} = 0$ holds along the trajectory of the test particle being considered. Therefore a restriction for the region in which the motion takes place is given by

$$0 \le e^{2(U-k)} (p_\varrho^2 + p_\zeta^2) = \left[\frac{1}{\varrho^2} (L^2 g_{tt} + 2LE g_{\varphi t} + E^2 g_{\varphi\varphi}) - 1 \right]. \quad (2.302)$$

---

[22] For a more general presentation of the associated Hamiltonian mechanics, see Ansorg (1998). The stochastic behaviour of the geodesics turns out to be related to the position of the region containing all the crossing points of the particle through the plane of the disc. If this region contains points lying inside the disc as well as points outside, the geodesic motion shows highly stochastic behaviour. However, if the crossing region is completely inside or outside the disc, the motion proves to be nearly integrable. In such cases, the corresponding Hamiltonian system is close to an integrable system of the so-called Liouville class, see e.g. Contopoulos (1994).

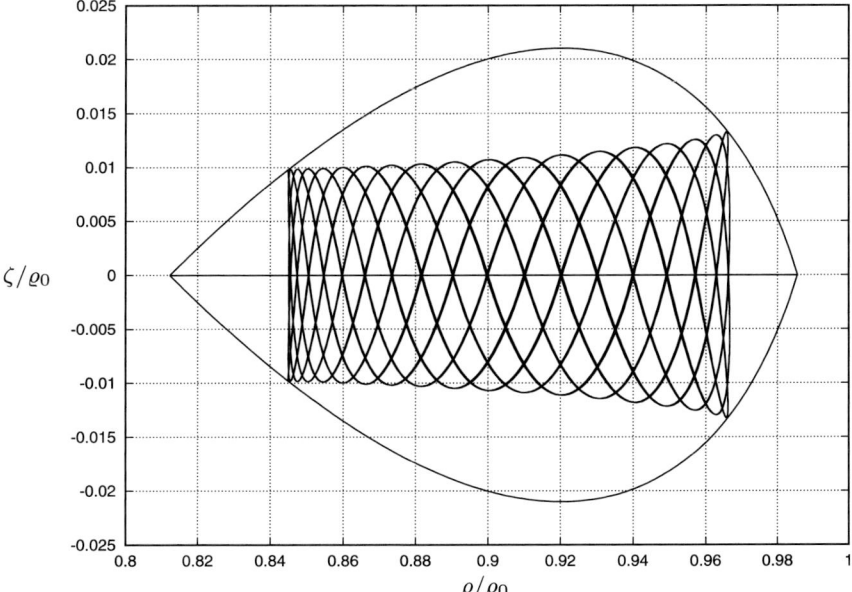

Fig. 2.19. Example of geodesic motion in the (ϱ, ζ)-space within the gravitational field of a disc with parameter $\mu = 3$ and radius ϱ_0. Specific angular momentum and energy are chosen to be $L = 2.78, E = 0.77$. The bounded area within which the motion takes place, characterized by Equation (2.302), is shown (after Ansorg 1998).

This region of motion shows the following properties (for an illustrative example see Fig. 2.19):

(i) The region is bounded for $E^2 < 1$. For $E^2 \geq 1$ there are unbounded areas allowing an escape of the test particle to infinity. However, even in the case $E^2 \geq 1$, there may be additional, bounded regions in which the particle is trapped.
(ii) Particles with $L \neq 0$ cannot cross the rotational axis.
(iii) The momenta p_ϱ and p_ζ as well as the velocities $\dot{\varrho}$ and $\dot{\zeta}$ vanish at the rim of the region under consideration. Furthermore, at the boundary of the region, the acceleration vector $(\ddot{\varrho}, \ddot{\zeta})$ is perpendicular to the curve of the boundary and points inwards.
(iv) Because of the property $g_{ij}(\varrho, -\zeta) = g_{ij}(\varrho, \zeta)$, the regions of motion are symmetric with respect to reflection through the plane of the disc.

Motions within the plane of the disc are described by a Hamiltonian with merely one degree of freedom. Due to the conditions $\dot{\varrho} = 0 = \ddot{\varrho}$, circular orbits with centres on the rotation axis are fixed. These lead to the equations

$$L^2 g_{tt}(\varrho, 0) + 2LE g_{\varphi t}(\varrho, 0) + E^2 g_{\varphi\varphi}(\varrho, 0) = \varrho^2 \qquad (2.303)$$

and

$$L^2 g_{tt,\varrho}(\varrho,0) + 2LE g_{\varphi t,\varrho}(\varrho,0) + E^2 g_{\varphi\varphi,\varrho}(\varrho,0) = 2\varrho, \tag{2.304}$$

which serve to determine the parameter functions $L_\pm = L_\pm(\varrho)$ and $E_\pm = E_\pm(\varrho)$. For a circular orbit of given radius ϱ, these functions yield the associated pairs of angular momentum and energy. In general, there are two such pairs, denoted by $(L_+(\varrho), E_+(\varrho))$ and $(L_-(\varrho), E_-(\varrho))$ where (L_+, E_+) refers to an orbit of positive angular momentum and similarly (L_-, E_-) to an orbit with negative angular momentum. It turns out that the functions $L_+(\varrho)$ and $E_+(\varrho)$ exist for an arbitrary choice of $\varrho \geq 0$. However, for sufficiently large values of the parameter μ, there are no corresponding (L_-, E_-)-pairs for ϱ-values within a certain interval $(\tilde{\varrho}_1, \tilde{\varrho}_2)$.

In Figs 2.20–2.24, the functions $L_\pm = L_\pm(\varrho)$ and $E_\pm = E_\pm(\varrho)$ are displayed for different values of the parameter μ. A discussion of these pictures follows.

(i) The pictures for (L_+, E_+) are similar for all values of μ, see Fig. 2.20 for a particular example. There is a monotonic growth of the functions L_+ and E_+ for $\varrho < \varrho_0$, where ϱ_0 is the radius of the disc. At $\varrho = \varrho_0$, where $E_+ = 1$, both functions turn back and decrease monotonically until they reach another turning point at $\varrho = \varrho'$. For still greater values of the parameter ϱ, the functions L_+ and E_+ grow monotonically again.

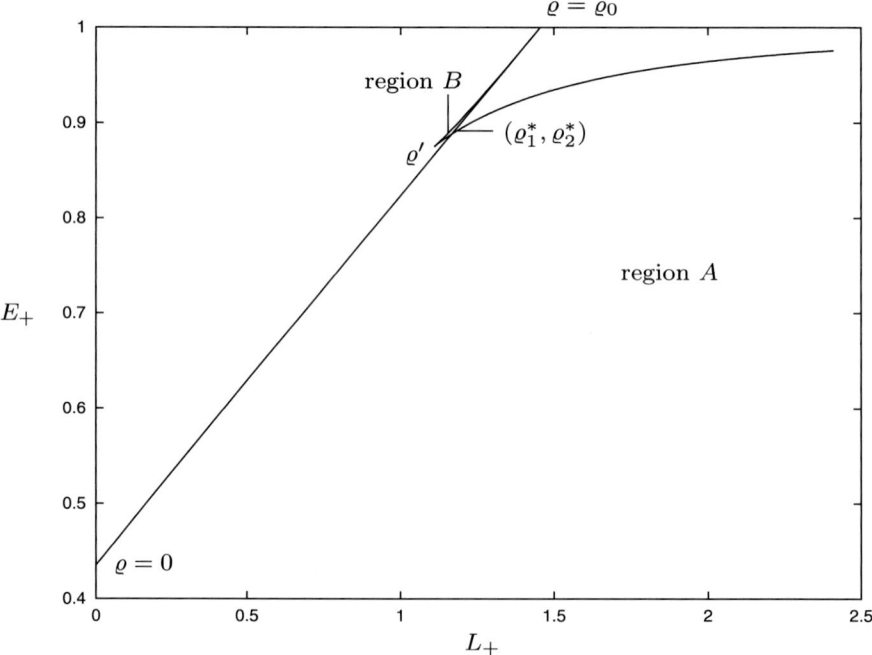

Fig. 2.20. For prograde circular orbits, the corresponding parameter curve $(L_+(\varrho), E_+(\varrho))$ is displayed for a disc with $\mu = 1.61$ (after Ansorg 1998).

2.3 The rigidly rotating disc of dust

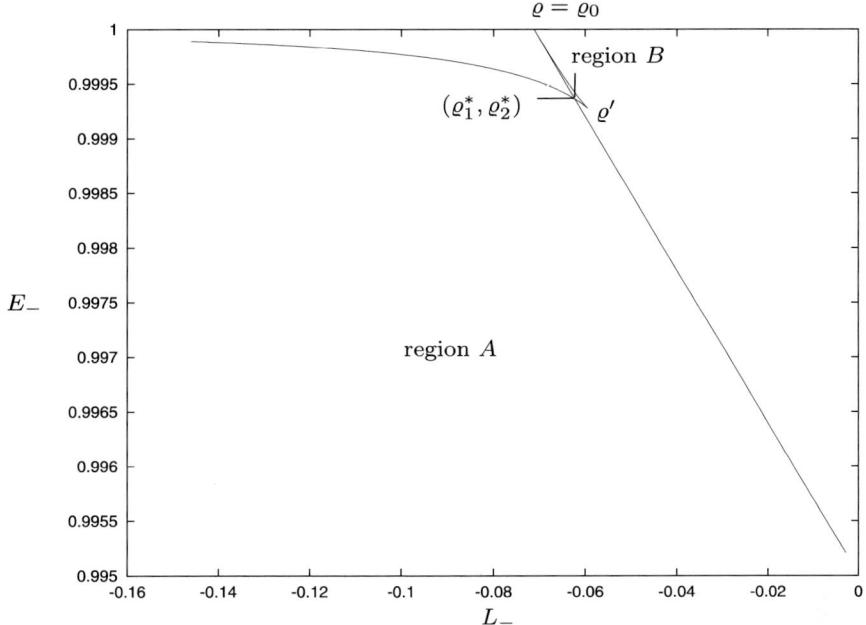

Fig. 2.21. For retrograde circular orbits, the corresponding parameter curve $(L_-(\varrho), E_-(\varrho))$ is displayed for a disc with $\mu = 0.01$ (after Ansorg 1998).

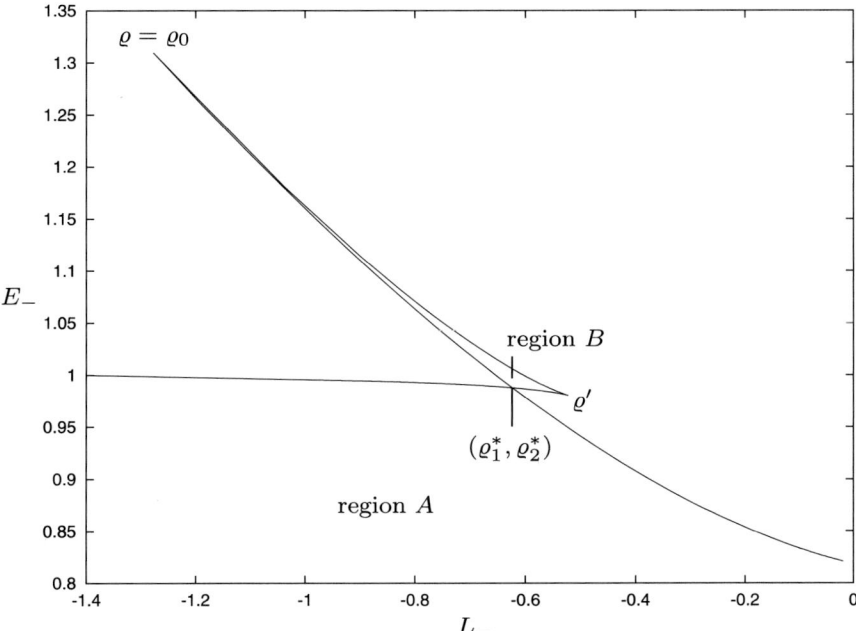

Fig. 2.22. For retrograde circular orbits, the corresponding parameter curve $(L_-(\varrho), E_-(\varrho))$ is displayed for a disc with $\mu = 0.4$ (after Ansorg 1998).

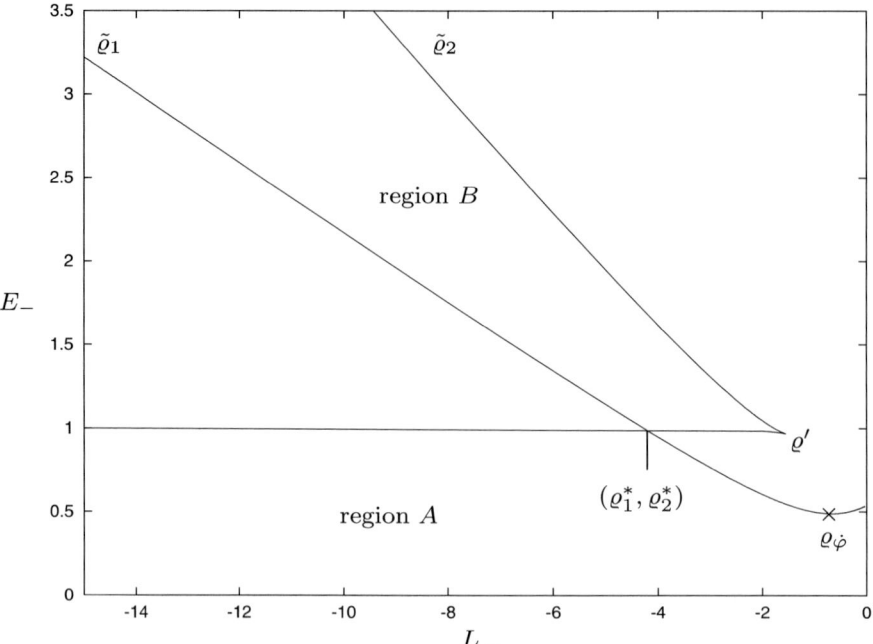

Fig. 2.23. For retrograde circular orbits, the corresponding parameter curve $(L_-(\varrho), E_-(\varrho))$ is displayed for a disc with $\mu = 1.21$ (after Ansorg 1998).

The turning points are characterized by the conditions $dL_+/d\varrho = 0 = dE_+/d\varrho$. As the radius $\varrho \to \infty$, E_+ tends to 1 and L_+ to $+\infty$. Circular orbits with a radius $\varrho \in (\varrho_0, \varrho')$ are unstable. Stable circular orbits are those with a radius $\varrho \in [0, \varrho_0)$ or $\varrho \in (\varrho', \infty)$; the remaining circular orbits with radius ϱ_0 or ϱ' are marginally unstable.

For circular orbits, the condition

$$\dot{\varphi} = \frac{dE}{dL} \dot{t} \qquad (2.305)$$

holds. In all sections of the graphs of the functions L_+ and E_+, the function $E_+ = E_+(L_+)$ grows monotonically. Hence, all circular orbits with positive angular momenta possess positive angular velocities. Furthermore, circular orbits with radii $\varrho < \varrho_0$ are precisely those paths along which the dust particles of the disc move. Their 4-velocity u^i is given by $(u^i) = (0, 0, \dot{\varphi}, \dot{t}) = e^{-V_0}(0, 0, \Omega, 1)$, cf. (1.18), (1.105), from which $dE/dL = \Omega = $ constant follows. Thus, the graph $E_+(L_+)$ is a straight line for $\varrho \in [0, \varrho_0]$.

The region A contains all (L, E)-pairs for which no motion of a particle is possible, i.e. either the restriction (2.302) cannot be met or the motion possesses a negative $\dot{t}$. If one chooses a (not necessarily circular) motion corresponding to an (L, E)-pair inside the small region B, then there exist two separate compact regions in which bounded motions are possible. At the intersection point of the stable and unstable parts of the

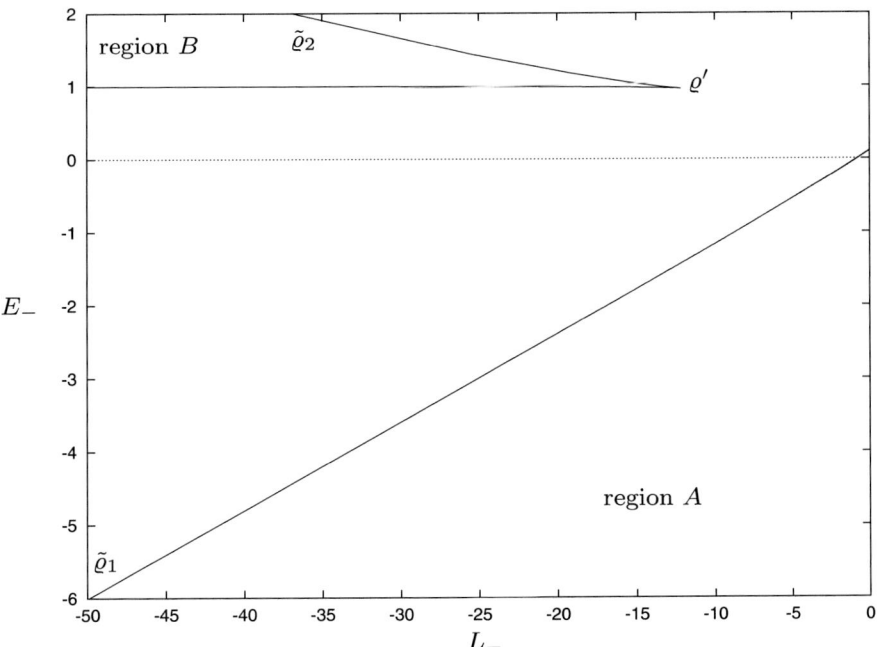

Fig. 2.24. For retrograde circular orbits, the corresponding parameter curve $(L_-(\varrho), E_-(\varrho))$ is displayed for a disc with $\mu = 3.5$ (after Ansorg 1998).

(L_+, E_+)-graphs, these regions degenerate to become two separate stable circular orbits with radii $\varrho_1^* < \varrho_0$ and $\varrho_2^* > \varrho_0$.

(ii) In Figs 2.21–2.24, the parameter functions $L_-(\varrho)$ and $E_-(\varrho)$ are displayed for different values of the parameter μ. For $0 < \mu < 1/2$, the pictures are similar to the graphs of the functions L_+ and E_+. The region B now extends to E_--values greater than 1. Thus, there are stable as well as unstable circular orbits and, furthermore, extended compact regions of motion with energies greater than 1 and corresponding negative angular momenta. From Equation (2.305) and the slopes of the functions $E_- = E_-(L_-)$, see Figs 2.21 and 2.22 for two examples, one concludes that all particles moving along circular orbits with negative angular momenta have negative angular velocities.

(iii) As μ approaches the value $\frac{1}{2}$, the region B grows and is unbounded for $\mu \geq \frac{1}{2}$. For radii $\varrho \in (\tilde{\varrho}_1, \tilde{\varrho}_2)$ there are no circular orbits with negative angular momentum.

As can be seen in Fig. 2.23, for sufficiently large values of μ ($\mu > \mu_{\dot\varphi} \approx 0.7088$) there are small intervals $[0, \varrho_{\dot\varphi})$ in which the functions $E_-(L_-)$ grow monotonically. Hence, circular orbits with radii $\varrho < \varrho_{\dot\varphi}$ possess positive angular velocities in spite of their having negative angular momenta, cf. page 171. At the radius $\varrho_{\dot\varphi}$, a particle may remain at rest (in the chosen coordinate system). Such a particle has the smallest energy possible of all motions around the disc.

(iv) Finally, in the range $\mu_e < \mu < \mu_0$, there are stable circular orbits with negative energies, see Fig. 2.24.[23] As the slope of $E_-(L_-)$ is positive for $\varrho < \tilde{\varrho}_1$, circular orbits with radii in this range have positive angular velocities.

(v) As $\mu \to 0$, the curves $L_\pm(\varrho)$ and $E_\pm(\varrho)$ tend to their corresponding counterparts for the Maclaurin disc. Since negative and positive angular momenta are equivalent in Newtonian theory, these graphs possess a reflectional symmetry with respect to the axis $L = 0$.

For the functions L_+ and E_+, the radii ϱ_1^*, ϱ_2^* and ϱ', as they depend on the parameter μ, can be seen in Fig. 2.25. Likewise, the μ-dependencies of the radii $\{\varrho_1^*, \varrho_2^*, \varrho', \tilde{\varrho}_1, \tilde{\varrho}_2, \varrho_{\dot\varphi}\}$ for the functions L_- and E_- are displayed in Fig. 2.26. In addition, the radii $\varrho_1^{(E)}$ and $\varrho_2^{(E)}$ at which the function $E_- = E_-(\varrho)$ reaches the value $E_-(\varrho_i^{(E)}) = 1$ can be seen.[24] For $\mu < \frac{1}{2}$, these radii are located in the

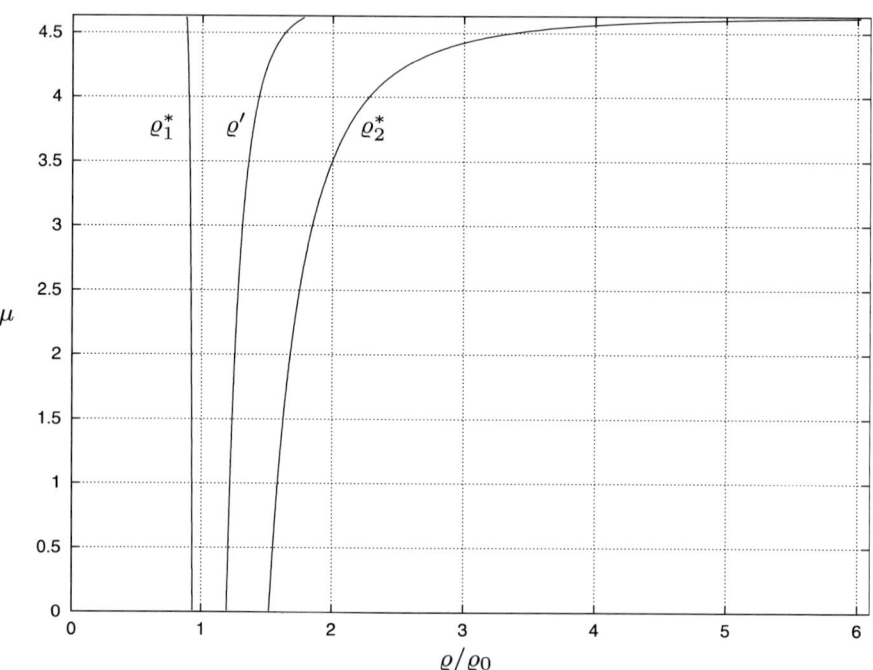

Fig. 2.25. The μ-dependency of the radii ϱ_1^*, ϱ_2^* and ϱ' for prograde circular orbits, described by the angular momentum-energy pair (L_+, E_+). For each disc parameter $\mu \in (0, \mu_0)$ there is a unique pair $(L_+^*(\mu), E_+^*(\mu))$ for which two separate stable circular orbits of radii ϱ_1^* and ϱ_2^* exist. Circular orbits with radii $\varrho \in (\varrho_0, \varrho')$, and only such orbits, are unstable (after Ansorg 1998).

---

[23] For $\mu > \mu_e \approx 1.69$, the disc is sufficiently relativistic so as to produce an ergosphere, cf. Fig. 2.15 and (2.283).
[24] Note that $E_-(\varrho') < 1$ always holds.

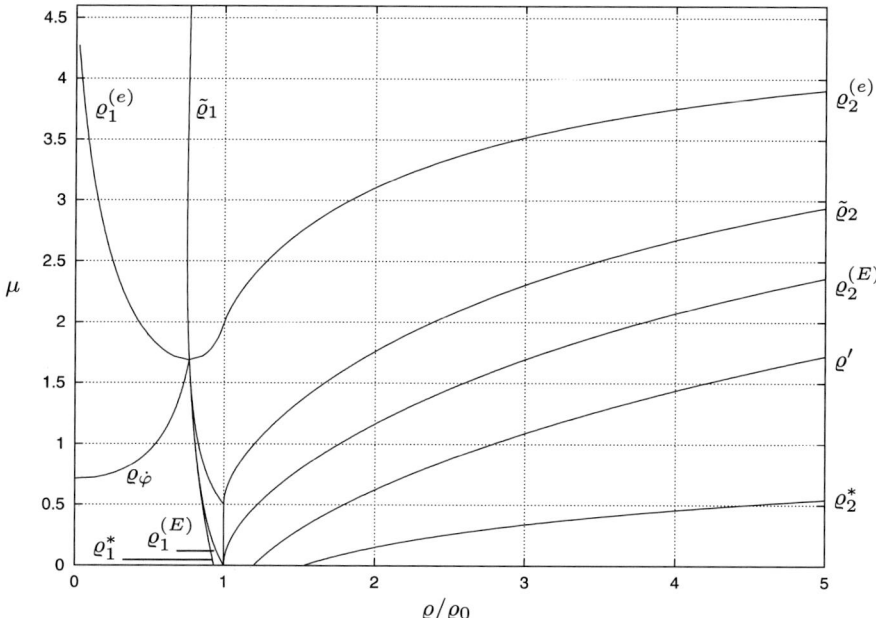

Fig. 2.26. The μ-dependency of the radii $\{\varrho_1^*, \varrho_2^*, \varrho', \tilde{\varrho}_1, \tilde{\varrho}_2, \varrho_{\dot\varphi}, \varrho_1^{(E)}, \varrho_2^{(E)}\}$ for retrograde circular orbits, described by the angular momentum-energy pair (L_-, E_-). For each disc parameter $\mu \in (0, \mu_e)$, there is a unique pair $(L_-^*(\mu), E_-^*(\mu))$ for which two separate stable circular orbits at the radii ϱ_1^* and ϱ_2^* exist. Circular orbits with radii $\varrho \in (\varrho_0, \varrho')$, and only such orbits, are unstable. For $\mu > \frac{1}{2}$, no circular orbits exist with radii $\varrho \in (\tilde{\varrho}_1, \tilde{\varrho}_2)$. At the radii $\varrho_{1/2}^{(E)}$, the corresponding circular orbit is characterized by the energy $E_- = 1$. For the circular orbit at the radius $\varrho_{\dot\varphi}$, the angular velocity $\dot\varphi$ vanishes, i.e. the test particle remains at rest (in the chosen coordinate system). Additionally, the radii $\varrho_1^{(e)}$ and $\varrho_2^{(e)}$ at which the ergosphere intersects the plane of the disc are plotted (after Ansorg 1998).

intervals

$$\varrho_1^* < \varrho_1^{(E)} < \varrho_0, \qquad \varrho_0 < \varrho_2^{(E)} < \varrho',$$

whereas for $\frac{1}{2} < \mu < \mu_e$,

$$\varrho_1^* < \varrho_1^{(E)} < \tilde{\varrho}_1, \qquad \tilde{\varrho}_2 < \varrho_2^{(E)} < \varrho'$$

hold. In the range $\mu_e < \mu < \mu_0$, the radius $\varrho_2^{(E)}$ is still within $(\tilde{\varrho}_2, \varrho')$ whereas the radii ϱ_1^*, ϱ_2^* and $\varrho_1^{(E)}$ cannot be defined since the curve $(E_-(\varrho), L_-(\varrho))$ does not possess an intersection point.

Circular orbits around a Kerr black hole are characterized by the functions, see e.g. Bardeen (1973b),

$$E_\pm = \frac{r^{3/2} - 2Mr^{1/2} \pm JM^{-1/2}}{r^{3/4}\sqrt{r^{3/2} - 3Mr^{1/2} \pm 2JM^{-1/2}}}, \qquad (2.306)$$

$$L_\pm = \pm\frac{M^{1/2}[r^2 \mp 2J\sqrt{r/M} + J^2M^{-2}]}{r^{3/4}\sqrt{r^{3/2} - 3Mr^{1/2} \pm 2JM^{-1/2}}}, \qquad (2.307)$$

where $\varrho = \sqrt{r^2 - 2Mr + J^2M^{-2}}$, M denotes the mass and J the angular momentum of the black hole (see Subsection 2.4). For a vanishing denominator in the above expressions, when $\varrho = \tilde{\varrho}$ say, $L_\pm \to \pm\infty$ and $E_\pm \to +\infty$. Circular orbits with radii smaller than $\tilde{\varrho}$ do not exist. For $\varrho > \tilde{\varrho}$, the $(L_\pm, E_\pm)$-curves are similar to those of Fig. 2.24 for $\varrho > \tilde{\varrho}_2$. Again there is a ϱ' that separates unstable (for $\varrho < \varrho'$) and stable orbits (for $\varrho > \varrho'$). Furthermore, for circular orbits within the Kerr metric, the signs of the angular momentum and angular velocity always coincide.

Generally, one finds that the qualitative behaviour of circular motions at sufficiently large radii is similar for the Kerr black hole and the disc. However, circular motions at small radii are quite different.

2.3.5 Black hole limit

It will turn out that $\varrho_0 \to 0$ in the black hole limit. Therefore, we first return to the non-normalized quantities $K, K_{a/b}$ and $K_{1/2}$, cf. (2.99), and introduce

$$H \equiv \frac{\mu}{\varrho_0^2}h, \quad Z_1 \equiv \varrho_0 W_1, \quad Z \equiv \frac{\varrho_0^3}{\mu}W \qquad (2.308)$$

together with

$$v_0 \equiv \frac{\mu}{\varrho_0^2}u, \quad v_1 \equiv \frac{\mu}{\varrho_0}v, \quad v_2 \equiv \mu w. \qquad (2.309)$$

With these expressions, the Ernst potential of the disc solution represented in the form (2.100), (2.101) reads

$$f = \exp\left\{\int_{K_1}^{K_a}\frac{K^2 dK}{Z} + \int_{K_2}^{K_b}\frac{K^2 dK}{Z} - v_2\right\} \qquad (2.310)$$

2.3 The rigidly rotating disc of dust

with

$$Z = \sqrt{(K + \mathrm{i}z)(K - \mathrm{i}\bar{z})(K^2 - K_1^2)(K^2 - K_2^2)}, \qquad (2.311)$$

$$K_1 = -\bar{K}_2 = \varrho_0 \sqrt{\frac{\mathrm{i} - \mu}{\mu}} \quad (\Re K_1 < 0). \qquad (2.312)$$

The upper limits of integration K_a and K_b in (2.310) have to be calculated from

$$\int_{K_1}^{K_a} \frac{\mathrm{d}K}{Z} + \int_{K_2}^{K_b} \frac{\mathrm{d}K}{Z} = v_0, \quad \int_{K_1}^{K_a} \frac{K\,\mathrm{d}K}{Z} + \int_{K_2}^{K_b} \frac{K\,\mathrm{d}K}{Z} = v_1 \qquad (2.313)$$

with

$$v_0 = \int_{-\mathrm{i}\varrho_0}^{+\mathrm{i}\varrho_0} \frac{H}{Z_1}\mathrm{d}K, \quad v_1 = \int_{-\mathrm{i}\varrho_0}^{+\mathrm{i}\varrho_0} \frac{H}{Z_1}K\,\mathrm{d}K, \quad v_2 = \int_{-\mathrm{i}\varrho_0}^{+\mathrm{i}\varrho_0} \frac{H}{Z_1}K^2\,\mathrm{d}K, \qquad (2.314)$$

$$H = \frac{\mu \ln\left[\sqrt{1 + \mu^2(1 + K^2/\varrho_0^2)^2} + \mu(1 + K^2/\varrho_0^2)\right]}{\pi\mathrm{i}\varrho_0^2\sqrt{1 + \mu^2(1 + K^2/\varrho_0^2)^2}} \quad (\Re H = 0), \qquad (2.315)$$

$$Z_1 = \sqrt{(K + \mathrm{i}z)(K - \mathrm{i}\bar{z})} \quad (\Re Z_1 < 0 \text{ for } \varrho, \zeta \text{ outside the disc}). \qquad (2.316)$$

The solution depends on the two parameters ϱ_0 and μ. The original parameters V_0 and Ω of the boundary value problem are related to ϱ_0 and μ as follows: $V_0 \equiv U(\varrho = 0, \zeta = 0)$ depends on μ alone. Using (2.118), it can be expressed as

$$V_0 = -\frac{1}{2}\mathrm{arcsinh}\left\{\mu + \frac{1 + \mu^2}{\wp[I(\mu); \frac{4}{3}\mu^2 - 4, \frac{8}{3}\mu(1 + \mu^2/9)] - \frac{2}{3}\mu}\right\}, \qquad (2.317)$$

$$I(\mu) = \frac{1}{\pi}\int_0^\mu \frac{\ln(x + \sqrt{1 + x^2})\,\mathrm{d}x}{\sqrt{(1 + x^2)(\mu - x)}} \qquad (2.318)$$

with the Weierstrass function $\wp$ defined by

$$\int_{\wp(x;g_2,g_3)}^{\infty} \frac{\mathrm{d}t}{\sqrt{4t^3 - g_2 t - g_3}} = x. \qquad (2.319)$$

Note that Equation (2.317) is equivalent to (2.201). The range $0 < \mu < \mu_0 = 4.62966184\ldots$ corresponds to $0 > V_0 > -\infty$ [μ_0 is the first zero of the denominator in (2.317)]. Recalling the definition of μ in (2.79),

$$\mu = 2\Omega^2 \varrho_0^2 e^{-2V_0}, \tag{2.320}$$

one can then find the relation $\Omega = \Omega(\mu, \varrho_0)$.

The limit $\mu \to \mu_0$

As was shown in Subsection 2.3.4, one obtains

$$\frac{M^2}{J} \to 1 \quad \text{and} \quad 2\Omega M \to 1 \tag{2.321}$$

in the limit $\mu \to \mu_0$, i.e. $V_0 \to -\infty$. From (2.320) we find

$$\Omega \varrho_0 \to 0, \tag{2.322}$$

which means that for finite M,

$$\varrho_0 \to 0. \tag{2.323}$$

Note that Ω remains a free parameter in the limit. For ease of discussion, we introduce spherical-like coordinates R, ϑ,

$$\varrho = R \sin \vartheta, \quad \zeta = R \cos \vartheta, \quad 0 \le \vartheta \le \pi. \tag{2.324}$$

The disc ($\zeta = 0, \varrho \le \varrho_0$) shrinks to the origin $R = 0$ of the coordinate system. For $R \ne 0$, the disc metric becomes exactly the $r > M$ part of the extreme Kerr metric with the radial Boyer–Lindquist coordinate r related to R by

$$r = R + M. \tag{2.325}$$

This can be shown as follows (Meinel 2002). Let us first rewrite (2.310) and (2.313) in the equivalent form

$$f = \exp\left\{\int_{K_b}^{K_a} \frac{K^2 \mathrm{d}K}{Z} - \tilde{v}_2\right\}, \quad \int_{K_b}^{K_a} \frac{\mathrm{d}K}{Z} = \tilde{v}_0, \quad \int_{K_b}^{K_a} \frac{K \mathrm{d}K}{Z} = \tilde{v}_1, \tag{2.326}$$

with

$$\tilde{v}_n = v_n - \int_{K_1}^{K_2} \frac{K^n \mathrm{d}K}{Z} \quad (n = 0, 1, 2). \tag{2.327}$$

(K_b is now on the other sheet of the Riemann surface.) In the limit $\mu \to \mu_0$, one obtains for $R > 0$, using (2.320) and (2.317),

$$\tilde{v}_0 = \frac{2\Omega}{R} - \frac{\pi \mathrm{i} \cos \vartheta}{2R^2}, \quad \tilde{v}_1 = -\frac{\pi \mathrm{i}}{2R}, \quad \tilde{v}_2 = 0 \tag{2.328}$$

2.3 The rigidly rotating disc of dust

(modulo periods). In the above integrals from K_b to K_a, Z can be replaced by $Z = K^2 \sqrt{(K + \mathrm{i}z)(K - \mathrm{i}\bar{z})}$ since K_1 and K_2 both tend to zero, cf. (2.312). Hence, all integrals become elementary and the unique result is

$$f = \frac{2\Omega R - 1 - \mathrm{i}\cos\vartheta}{2\Omega R + 1 - \mathrm{i}\cos\vartheta} \quad (R > 0), \quad (2.329)$$

which is the Ernst potential of the extreme Kerr solution. Note that $R = 0$ ($r = M$) characterizes the horizon (and the throat) of the extreme Kerr black hole. $\Omega = \Omega_\mathrm{h} = 1/(2M)$ is the 'angular velocity of the horizon', see (1.149).

A completely different limit of the spacetime for $\mu \to \mu_0$ is obtained for finite values of R/ϱ_0 (corresponding to the previously excluded $R = 0$). Therefore, we consider a coordinate transformation (Bardeen and Wagoner 1971, Meinel 2002)

$$\tilde{r} = e^{-V_0} R, \quad \tilde{\vartheta} = \vartheta, \quad \tilde{\varphi} = \varphi - \Omega t, \quad \tilde{t} = e^{V_0} t. \quad (2.330)$$

Note that finite R/ϱ_0 correspond to finite $\tilde{r}$ in the limit, as can be seen from (2.320). For $\mu < \mu_0$, this is nothing other than the transformation to the corotating system combined with a rescaling of R and t. The transformed Ernst potential $\tilde{f}$ is related to the Ernst potential f' in the corotating system ($R' = R$, $\vartheta' = \vartheta$, $\varphi' = \varphi - \Omega t$, $t' = t$) according to $\tilde{f} = f' \exp(-2V_0)$, i.e.

$$\frac{\tilde{f}}{\tilde{r}^2} = \frac{f'}{R^2} \quad \text{for} \quad \mu < \mu_0. \quad (2.331)$$

However, for $\mu \to \mu_0$, the solutions f' (finite $R > 0$) and $\tilde{f}$ (finite $\tilde{r}$) separate from each other.

For finite $R > 0$, the extreme Kerr solution arises,[25] while finite $\tilde{r}$ values lead to a solution that still describes a disc, whose finite coordinate radius is

$$\tilde{\varrho}_0 = \lim_{\mu \to \mu_0} (e^{-V_0} \varrho_0) = \sqrt{2\mu_0}\, M. \quad (2.332)$$

Note that the proper radius of the disc remains finite in the limit $\mu \to \mu_0$ as well. Its circumference is $4\pi M \sqrt{\mu_0/2 - 1}$, which is larger[26] than the circumference $4\pi M$ of the extreme Kerr throat. The metric corresponding to $\tilde{f}$, which can be expressed in terms of theta functions, is regular everywhere outside the disc, but it is not asymptotically flat. The spacetime structure of both solutions (f' and $\tilde{f}$) coincides at $R \to 0$ (the throat) and $\tilde{r} \to \infty$ (spatial infinity). The relation (2.331) survives in the form

$$\lim_{\tilde{r} \to \infty} \frac{\tilde{f}}{\tilde{r}^2} = \lim_{R \to 0} \frac{f'}{R^2} \quad \text{as} \quad \mu \to \mu_0. \quad (2.333)$$

---

[25] Accordingly, in the limit $\mu \to \mu_0$, all gravitational multipole moments assume the extreme Kerr values, cf. Subsection 2.3.4.
[26] An analogous situation will be found for fluid rings approaching the black hole limit, see Fig. 3.22.

[The limits have to be taken consistently with (2.330).] The Ernst potential f' of the extreme Kerr solution in the corotating system reads

$$f' = -\Omega^2 R^2 \left[\frac{2(1 + i\cos\vartheta)^2}{2\Omega R + 1 - i\cos\vartheta} + \sin^2\vartheta \right]. \qquad (2.334)$$

Accordingly, for $\mu = \mu_0$ and $\tilde{r} \to \infty$,

$$\tilde{f} \to \tilde{f}_{as} = -\Omega^2 \tilde{r}^2 \left[\frac{2(1 + i\cos\tilde{\vartheta})^2}{1 - i\cos\tilde{\vartheta}} + \sin^2\tilde{\vartheta} \right]. \qquad (2.335)$$

Note that $\tilde{f}_{as}$ belongs to the family of solutions to the Ernst equation of the type $f = R^k Y_k(\cos\vartheta)$ presented by Ernst (1977). The corresponding *asymptotic* line element is given by the 'extreme Kerr throat geometry' or 'near-horizon geometry' (1.150).

The analytic solution to the disc of dust problem thus provides an explicit example for a parametric transition from a normal matter equilibrium configuration to a black hole as discussed in Subsection 1.8.3. Further examples, obtained by numerical means, will be presented in Chapter 3.

2.4 The Kerr metric as the solution to a boundary value problem

In this section, we want to show that the 'inverse method' has another remarkable application: the derivation of the Kerr metric as the unique solution to a well-defined boundary value problem (Neugebauer 2000, Neugebauer and Meinel 2003).

To obtain the stationary and axisymmetric, asymptotically flat solution describing the vacuum exterior of a black hole, we have to solve the following boundary value problem of the Ernst equation (see Subsection 1.8.1 and Fig. 2.27): On the horizon

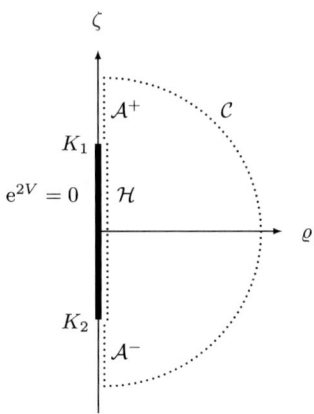

Fig. 2.27. The boundary value problem for a black hole, and the line of integration.

2.4 The Kerr metric as the solution to a boundary value problem

$\mathcal{H}$, which in canonical Weyl coordinates covers the domain

$$\mathcal{H}: \quad \varrho = 0, \quad K_1 \geq \zeta \geq K_2, \tag{2.336}$$

the real part e^{2V} of the Ernst potential f' in the corotating frame (rotating with the angular velocity $\Omega = \Omega_h$ of the horizon) has to vanish:[27]

$$\mathcal{H}: \quad e^{2V} \equiv -\chi^i \chi_i = 0, \tag{2.337}$$

whereas the Ernst potential f in the non-rotating frame has to be regular everywhere outside the horizon and to satisfy

$$f \to 1 \quad \text{as} \quad \varrho^2 + \zeta^2 \to \infty. \tag{2.338}$$

Because of

$$e^{2V} = e^{2U} \left[(1 + \Omega a)^2 - \Omega^2 \varrho^2 e^{-4U} \right], \tag{2.339}$$

cf. (2.238), this implies[28]

$$\mathcal{H}: \quad 1 + \Omega a = 0. \tag{2.340}$$

Remember that a vanishes on the regular parts of the axis,

$$\mathcal{A}^{\pm}: \quad a = 0, \tag{2.341}$$

i.e. a is not continuous at $\varrho = 0, \zeta = K_{1/2}$. However the metric coefficients, e.g. $g_{\varphi t} = -a e^{2U}$, are continuous ($e^{2U}$ vanishes at $\varrho = 0, \zeta = K_{1/2}$). It should be mentioned in this connection, that our derivation of the Kerr solution makes use of the assumption $\Omega \neq 0$. However, the final result will contain the Schwarzschild solution (corresponding to $\Omega = 0$) as a limiting case.

As a first step, we calculate Φ and Φ' along the horizon $\mathcal{H}$. From (2.41), (2.336), (2.337), (2.340) and (2.56) we obtain

$$\mathcal{H}: \quad \begin{aligned} \Phi &= \begin{pmatrix} \overline{f(\zeta)} & 1 \\ f(\zeta) & -1 \end{pmatrix} \begin{pmatrix} U(K) & V(K) \\ W(K) & X(K) \end{pmatrix}, \\ \Phi' &= 2i\Omega(K - \zeta) \begin{pmatrix} -1 & 0 \\ 1 & 0 \end{pmatrix} \begin{pmatrix} U(K) & V(K) \\ W(K) & X(K) \end{pmatrix}. \end{aligned} \tag{2.342}$$

The Ernst equations have to hold at K_1 and K_2 too. Hence, Φ and Φ' must be continuous at K_1 and K_2. Considering (2.57), (2.58), (2.59), (2.60) and (2.342), we

---

[27] It turns out that the imaginary part of f' (up to an arbitrary constant) vanishes on the horizon as well.
[28] Note that $e^{2U} \neq 0$ holds on the horizon of rotating black holes except at the poles, since $e^{2U} = -\Omega^2 \eta^i \eta_i$ follows from (1.125)–(1.127).

are led to the conditions

$$\begin{pmatrix} f_1 & -1 \\ f_1 + 2i\Omega(K - K_1) & -1 \end{pmatrix} \begin{pmatrix} F & 0 \\ G & 1 \end{pmatrix} =$$

$$\begin{pmatrix} f_1 & -1 \\ 2i\Omega(K - K_1) & 0 \end{pmatrix} \begin{pmatrix} U & V \\ W & X \end{pmatrix},$$

$$\begin{pmatrix} f_2 & -1 \\ f_2 + 2i\Omega(K - K_2) & -1 \end{pmatrix} \begin{pmatrix} 1 & G \\ 0 & F \end{pmatrix} =$$

$$\begin{pmatrix} f_2 & -1 \\ 2i\Omega(K - K_2) & 0 \end{pmatrix} \begin{pmatrix} U & V \\ W & X \end{pmatrix},$$

(2.343)

where $f_1 = f(\zeta = K_1)$ and $f_2 = f(\zeta = K_2)$. Note that f_1 and f_2 are purely imaginary. Eliminating the $UVWX$ matrix, we obtain

$$\mathcal{N} \equiv \begin{pmatrix} F & -G \\ G & (1 - G^2)/F \end{pmatrix} \qquad (2.344)$$

$$= \left(1 + \frac{\mathbf{F}_1}{2i\Omega(K - K_1)}\right)\left(1 + \frac{\mathbf{F}_2}{2i\Omega(K - K_2)}\right), \qquad (2.345)$$

where

$$\mathbf{F}_1 = \begin{pmatrix} -f_1 & 1 \\ -f_1^2 & f_1 \end{pmatrix}, \quad \mathbf{F}_2 = \begin{pmatrix} f_2 & -1 \\ f_2^2 & -f_2 \end{pmatrix}. \qquad (2.346)$$

Note that the matrix $\mathcal{N}$ is related to the previously introduced matrix $\mathcal{M}$, see (2.72), by $\mathcal{M} = \begin{pmatrix} 0 & -1 \\ 1 & 0 \end{pmatrix} \mathcal{N} \begin{pmatrix} -1 & 0 \\ 0 & 1 \end{pmatrix}$. Obviously, the elements of $\mathcal{N}$ are regular everywhere in the complex K-plane with the exception of the two simple poles at K_1 and K_2 ($\Im K_1 = 0 = \Im K_2$). The sum of the off-diagonal elements in (2.344) must be zero. This requirement leads to the constraints

$$f_1 = -f_2, \qquad \Omega = \frac{if_1(1 + f_1^2)}{(K_1 - K_2)(1 - f_1^2)}. \qquad (2.347)$$

$F(K)$ and $G(K)$ take the form

$$F(K) = \frac{4\Omega^2(K^2 - K_1^2) + 4i\Omega f_1 K - 2f_1^2}{4\Omega^2(K^2 - K_1^2)}, \qquad (2.348)$$

$$G(K) = \frac{4i\Omega K_1 + 2f_1}{4\Omega^2(K^2 - K_1^2)}. \qquad (2.349)$$

2.4 The Kerr metric as the solution to a boundary value problem

Here, we have chosen $K_1 = -K_2$, i.e. we have set the horizon in a symmetric position in the ϱ, ζ-plane, see Fig. 2.27. Making use of (2.63) and (2.64) and eliminating Ω by the second constraint equation, we obtain the axis potential

$$\mathcal{A}^+: \quad f = \frac{\zeta(1+f_1^2) + (f_1^2 - 1 + 2f_1)K_1}{\zeta(1+f_1^2) + (1 - f_1^2 + 2f_1)K_1}. \tag{2.350}$$

It is useful to introduce the multipole moments mass M and angular momentum J by an asymptotic expansion of f,

$$M = \frac{1 - f_1^2}{1 + f_1^2} K_1, \qquad \frac{J}{M} = \frac{2i f_1 K_1}{1 + f_1^2} \tag{2.351}$$

and to replace f_1 and K_1 in (2.348), (2.349), (2.350) and (2.347):

$$F(K) = \frac{(K+M)^2 + (J/M)^2}{K^2 + (J/M)^2 - M^2}, \qquad G(K) = \frac{-2i J}{K^2 + (J/M)^2 - M^2}. \tag{2.352}$$

To represent $f(\zeta)$, a simplifying parameterization is advisable,

$$f_1 = i \tan \delta/2, \quad J = -M^2 \sin \delta, \tag{2.353}$$

$$K_1 = -K_2 = \sqrt{M^2 - (J/M)^2} = M \cos \delta, \tag{2.354}$$

with the real parameter δ satisfying

$$-\frac{\pi}{2} < \delta < 0. \tag{2.355}$$

Note that we have taken into account the necessary condition $M > 0$. Moreover, we have chosen $J > 0$ (and thus $\Omega > 0$) without loss of generality.[29] This yields

$$\mathcal{A}^+: \quad f = \frac{\zeta - M + i M \sin \delta}{\zeta + M + i M \sin \delta}. \tag{2.356}$$

Finally, the second constraint equation (2.347) becomes the well-known relation,

$$2M\Omega = \frac{M^2}{J} - \sqrt{\frac{M^4}{J^2} - 1}, \tag{2.357}$$

connecting the angular velocity of the horizon with mass and angular momentum, cf. (1.136).

---

[29] The solution with negative J (and negative Ω) is simply given by the complex conjugate Ernst potential. Solutions with negative M are not singularity-free outside the horizon.

Solutions for which the Ernst potential along the axis is a rational function of ζ can uniquely be continued to all space using 'Bäcklund techniques', see Neugebauer (1996). The continuation of (2.356) gives

$$f(\varrho,\zeta) = \frac{r_1 e^{-i\delta} + r_2 e^{i\delta} - 2M\cos\delta}{r_1 e^{-i\delta} + r_2 e^{i\delta} + 2M\cos\delta}, \qquad (2.358)$$

where the non-negative quantities r_i are defined by

$$r_i^2 = (K_i - \zeta)^2 + \varrho^2 \qquad (i = 1, 2) \qquad (2.359)$$

with K_i as in (2.354). This is the Ernst potential f of the Kerr solution[30] in Weyl–Lewis–Papapetrou coordinates including the limiting cases $\delta \to 0$ (Schwarzschild solution, $J = 0$) and $\delta \to -\pi/2$ (extreme Kerr solution, $J = M^2$). To obtain the latter limit one has to apply the Bernoulli–l'Hospital rule. The extreme Kerr black hole is characterized by a degenerate horizon (vanishing 'surface gravity'). The black hole uniqueness theorems, see Heusler (1996), state that the Kerr black holes with $J < M^2$ are the only stationary vacuum black holes with non-degenerate horizon. It is an interesting question whether the extreme Kerr black hole is the only stationary vacuum black hole with degenerate horizon. At least if axisymmetry and the corresponding conditions (1.125), (1.126) are assumed to hold from the very beginning, this question can be answered in the affirmative – as will be shown in the remainder of this section.

The degenerate case

The extreme Kerr solution results from the limit $\delta \to -\pi/2$, i.e. the interval $K_1 \geq \zeta \geq K_2$ of the ζ-axis characterizing the horizon in canonical Weyl coordinates shrinks to a point ($K_1 = 0 = K_2$). We now want to show that the extreme Kerr solution follows uniquely, if we *start* with the assumption that the horizon is located at a point on the ζ-axis. Of course, the slice $t =$ constant, $\varphi =$ constant of the horizon still has to be one-dimensional. Therefore, we have to parameterize the position along the horizon by another coordinate. A natural choice is the angle ϑ of spherical-like coordinates R, ϑ:

$$\varrho = R\sin\vartheta, \quad \zeta = R\cos\vartheta, \quad 0 \leq \vartheta \leq \pi, \qquad (2.360)$$

cf. (2.324). The horizon is placed at the origin ($R = 0$) of our Weyl-coordinate system, see Fig. 2.28. The 'north pole' of the horizon is at $\vartheta = 0$, the 'south pole'

---

[30] The full metric is given by Equation (1.130). It can be calculated from the Ernst potential (2.358) using (1.44), (1.41) and (1.40).

2.4 The Kerr metric as the solution to a boundary value problem

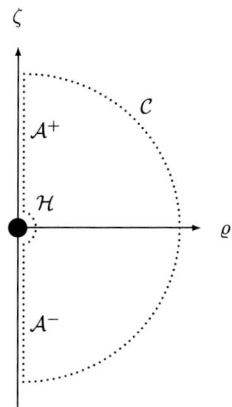

Fig. 2.28. The black hole with degenerate horizon ($\varrho = \zeta = 0$).

at $\vartheta = \pi$. The calculation of Φ and Φ' along the horizon gives

$$\mathcal{H}: \quad \begin{aligned} \Phi &= \begin{pmatrix} \overline{f(\vartheta)} & 1 \\ f(\vartheta) & -1 \end{pmatrix} \begin{pmatrix} U(K) & V(K) \\ W(K) & X(K) \end{pmatrix}, \\ \Phi' &= 2\mathrm{i}\Omega K \begin{pmatrix} -1 & 0 \\ 1 & 0 \end{pmatrix} \begin{pmatrix} U(K) & V(K) \\ W(K) & X(K) \end{pmatrix}. \end{aligned} \quad (2.361)$$

We can now repeat our analysis of the continuity of Φ and Φ' at the 'poles' of the horizon, which touch the regular parts of the axis $\mathcal{A}^\pm$. The unique (positive mass) result for the axis Ernst potential is

$$\mathcal{A}^+: \quad f = \frac{\zeta - M - \mathrm{i}M}{\zeta + M - \mathrm{i}M}, \quad M = \frac{1}{2\Omega}, \quad (2.362)$$

where we have again assumed $\Omega > 0$. This is the axis potential of the extreme Kerr solution. The continuation to all space is given by

$$f = \frac{2\Omega R - 1 - \mathrm{i}\cos\vartheta}{2\Omega R + 1 - \mathrm{i}\cos\vartheta}, \quad (2.363)$$

cf. (2.329).

Since a finite interval and a single point on the ζ-axis are the only possibilities for the (Killing-) horizon, we can conclude that the Kerr black holes – including the extreme case – are the only stationary and axisymmetric black holes surrounded by a vacuum.

3
Numerical treatment of the general case

After introducing basic notions and equations concerning fluid bodies in the first chapter, we turned our attention to the analytical treatment of a small number of limiting cases in the second. The methods presented there are quite powerful, as was shown, but also have severe limitations. In particular, it does not seem to be possible to use them for treating genuinely three-dimensional fluid bodies in which the shape of the free boundary emerges out of the simultaneous solution to an interior and an exterior problem along with transition conditions. We thus devote the third chapter to a numerical treatment of the problem, which is capable of handling the general case.

The form of the metric we use in this chapter is not (1.5) as with the disc, but rather (1.6)

$$ds^2 = e^{2\alpha}(d\varrho^2 + d\zeta^2) + W^2 e^{-2\nu}(d\varphi - \omega\, dt)^2 - e^{2\nu} dt^2. \tag{3.1}$$

The benefit of this line element in the context of numerical solutions is that ν remains real even within an ergosphere (see the related comment on page 14). The field equations are, cf. (1.29) and (1.33),

$$\nabla \cdot (B\nabla \nu) - \frac{1}{2}\varrho^2 B^3 e^{-4\nu}(\nabla \omega)^2 = 4\pi e^{2\alpha} B\left[(\epsilon + p)\frac{1+v^2}{1-v^2} + 2p\right], \tag{3.2a}$$

$$\nabla \cdot (\varrho^2 B^3 e^{-4\nu}\nabla \omega) = -16\pi \varrho B^2 e^{2\alpha - 2\nu}(\epsilon + p)\frac{v}{1-v^2}, \tag{3.2b}$$

$$\nabla \cdot (\varrho \nabla B) = 16\pi \varrho B e^{2\alpha} p \tag{3.2c}$$

and

$$\Delta_2 \alpha - \frac{1}{\varrho}\frac{\partial \nu}{\partial \varrho} + \nabla \nu \nabla u - \frac{1}{4}\varrho^2 B^2 e^{-4\nu}(\nabla \omega)^2 = -4\pi e^{2\alpha}(\epsilon + p) \tag{3.2d}$$

with

$$B := W/\varrho, \quad e^u := e^v/B \quad \text{and} \quad v := \varrho B e^{-2v}(\Omega - \omega).$$

The operator ∇ has the same meaning as in a Euclidean 3-space in which ϱ, ζ and φ are cylindrical coordinates. Thus the first three of the field equations can be applied as they are in other, related coordinates such as r, θ, φ with $\varrho = r \sin \theta$ and $\zeta = r \cos \theta$. In (3.2d), however, the operator $\triangle_2 := \partial^2/\partial \varrho^2 + \partial^2/\partial \zeta^2$ is not coordinate independent.

The vanishing divergence of the energy-momentum tensor yields a relationship between the specific enthalpy h and the metric functions given by (1.26) and (1.31),

$$h = h(0)\, e^{V_0 - V}, \qquad e^V = e^v \sqrt{1 - v^2}, \tag{3.3}$$

which allows us to express $\epsilon(h)$ and $p(h)$ on the right hand sides of Equations (3.2) as functions of $e^V = e^v \sqrt{1 - v^2}$. As was already discussed in Section 1.4, an alternative to the second order differential equation (3.2d) for determining α, is a line integral. Here we choose nonetheless to treat α on the same footing as the other functions since it renders the numerical methods more transparent without requiring prohibitively more computational resources.

This chapter deals with numerical solutions to the above field equations together with appropriate boundary, regularity and asymptotic conditions. The first section provides, by way of simple examples, relevant specifics of the numerical methods to be employed. The second provides a list of the coordinate mappings used in various scenarios. In the remaining sections, the physical properties of selected numerical solutions are presented. These include a detailed look at homogeneous figures, a discussion of a selection of further equations of state and finally a treatment of a central black hole surrounded by a ring.

3.1 A multi-domain spectral method

Numerical methods for solving the system of equations (3.2) describing the equilibrium of rotating relativistic stars have been developed since the 1970s (Wilson 1972, Bonazzola and Schneider 1974, Butterworth and Ipser 1976). The end of the 1980s and the beginning of the 1990s have seen a considerable further development and refinement of the techniques being used (Friedman *et al.* 1986, 1989, Komatsu *et al.* 1989a,b, Lattimer *et al.* 1990, Cook *et al.* 1992, Neugebauer and Herold 1992, Stergioulas and Friedman 1995; for a review see Stergioulas 2003).

The application of pseudo-spectral methods within the realm of numerical relativity was initiated by Bonazzola *et al.* (1993). An improved version of the corresponding code by Bonazzola et al. (1998) demonstrated the excellent convergence properties of pseudo-spectral methods, which can lead to extremely high accuracy.

In this section, we describe the pseudo-spectral method that was developed for the calculation of the numerical results presented in the remainder of this chapter (Ansorg et al. 2002a, 2003a). A particular feature of this method is the compactification of the entire spatial domain and the division into several subdomains, where one of the domain boundaries is chosen to coincide exactly with the surface of the fluid configuration. In this manner, it is possible to avoid the so-called Gibbs phenomenon entirely, which would affect the spectral convergence rate. As a starting point, we introduce and illustrate pseudo-spectral methods for ordinary differential equations. We then present coordinate mappings for each domain. The collection of field equations, boundary and transition conditions in each subdomain provides a high-dimensional algebraic system of equations for the potential values at specified discrete spatial gridpoints. We describe the structure of the solution vector from the Newton–Raphson scheme utilized in order to tackle this system of equations. The final result is a highly accurate spectral approximation of the desired gravitational field of rotating fluid configurations in equilibrium.

3.1.1 Chebyshev expansions

For a spectral expansion of a real-valued function f defined on the interval $[a, b]$ ($a, b \in \mathbb{R}$), we write

$$f(x) = \sum_{k=0}^{n-1} c_k^{(n)} \Phi_k(x) + R^{(n)}(x). \tag{3.4}$$

Here, the integer n is the spectral expansion order, $c_k^{(n)}$ are the spectral coefficients (which depend on the order n in general) and $R^{(n)}(x)$ is the residual term. The spectral basis functions Φ_k are chosen according to the properties of the underlying function f (e.g. trigonometric functions for periodic f). In this book, we use Chebyshev polynomials of the first kind defined on the interval $[a, b] = [0, 1]$, i.e.

$$\Phi_k(x) = T_k(2x - 1) \tag{3.5}$$

with

$$T_k(\xi) = \cos(k \arccos \xi). \tag{3.6}$$

Moreover we choose the residual $R^{(n)}$ to vanish at the spectral gridpoints

$$x_j^{(n)} = \sin^2\left(\frac{\pi j}{2(n-1)}\right); \quad j = 0, 1, \ldots, n-1; \qquad (3.7)$$

which implies

$$c_k^{(n)} = (-1)^k \frac{2 - \delta_{k,0} - \delta_{k,n-1}}{n-1}$$
$$\times \left(\frac{1}{2}\left[f(0) + (-1)^k f(1)\right] + \sum_{j=1}^{n-2} f(x_j^{(n)}) \cos\frac{\pi k j}{n-1}\right), \qquad (3.8)$$

where $\delta_{k,j}$ is the Kronecker symbol. Note that

$$x_0^{(n)} = 0, \quad x_{n-1}^{(n)} = 1 \qquad (3.9)$$

and

$$\Phi_{n-1}(x_j^{(n)}) = (-1)^{n-1-j}, \qquad (3.10)$$

i.e. the $x_j^{(n)}$ are the abscissa values of the (local) extrema of Φ_{n-1} in $[0, 1]$.

The reason for this choice is an extremely rapid fall-off of the coefficients $c_k^{(n)}$ provided that the function f is analytic within $[0, 1]$ and that there is no strong growth of the function's derivative values as one goes to higher and higher derivatives of f. In a certain sense, the Chebyshev expansion results in a polynomial representation that is close to the polynomial of best approximation (Arnold 2001). In the following we illustrate the behaviour of the spectral coefficients for some representative example functions.

1. Let us first consider

$$f_1(x) = \frac{1}{1+x^2} \qquad (3.11)$$

as an example of a smooth analytic function on $[0, 1]$. In Fig. 3.1, the absolute values of the coefficients $c_k^{(n)}$ for $n = 30$ are displayed (solid line). Moreover, the maximal residuals

$$R_{max}^{(n)} = \max_{x \in [0,1]} |R^{(n)}(x)| \qquad (3.12)$$

are plotted against the resolution orders n (dashed line). We see a rapid decrease in the magnitudes of the coefficients and also of the maximal residuals. The similarity

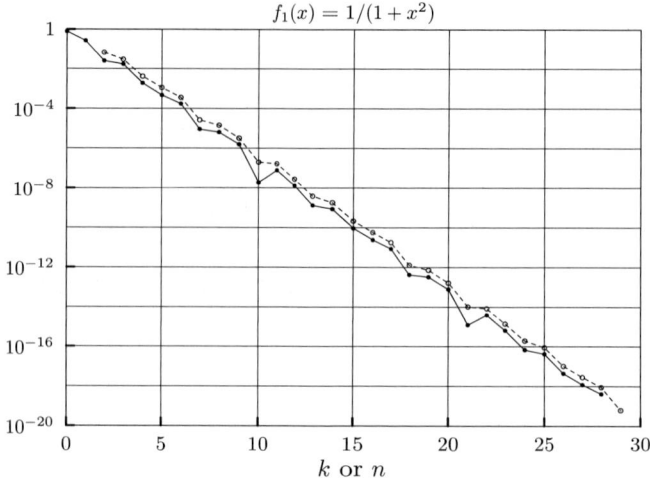

Fig. 3.1. Chebyshev coefficients $|c_k^{(30)}|$ (solid line) and maximal residuals $R_{\max}^{(n)}$ (dashed line) of the function f_1. The values along the abscissa are k for $|c_k^{(30)}|$ or alternatively n for $R_{\max}^{(n)}$.

between two such curves is a general feature of pseudo-spectral methods when applied to analytic functions f. In addition, in this semi-log plot the values oscillate slightly about a straight line, indicating an exponential convergence rate of the spectral expansions.

2. The next example demonstrates a situation that could model weak gravitational sources where steep gradients of the gravitational potentials are present. For $\epsilon > 0$ the function

$$f_2(x) = \frac{\epsilon}{\epsilon + x} \tag{3.13}$$

is analytic but the magnitudes of the derivatives become very large in the vicinity of $x = 0$ if $\epsilon \ll 1$, as revealed by the Taylor expansion

$$f_2(x) = \sum_{k=0}^{\infty} \left(-\frac{x}{\epsilon}\right)^k, \quad 0 \leq x < \epsilon. \tag{3.14}$$

The corresponding convergence rates of the spectral expansions become extremely slow as ϵ approaches zero, see Fig. 3.2.

3. The previous example begs the question as to the suitability of pseudo-spectral methods in describing the gravitational field around weak sources. In the third

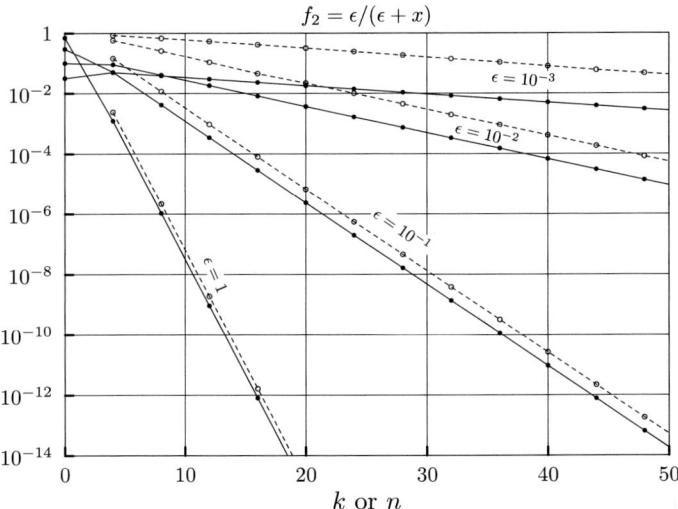

Fig. 3.2. Chebyshev coefficients $|c_k^{(100)}|$ (solid lines) and maximal residuals $R_{\max}^{(n)}$ (dashed lines) of the functions f_2 for several values of ϵ.

example we introduce the coordinate X,

$$x = \frac{\sinh(X \ln \epsilon)}{\sinh(\ln \epsilon)} = -2\epsilon \frac{\sinh(X \ln \epsilon)}{1 - \epsilon^2}, \qquad (3.15)$$

in which the derivatives of f_2 are merely of order $(\ln \epsilon)^k$ as opposed to ϵ^{-k}:

$$f_3(X) = f_2(x) = \frac{1 - \epsilon^2}{1 - \epsilon^2 - 2\sinh(X \ln \epsilon)}$$
$$= 1 + 2\frac{X \ln \epsilon}{1 - \epsilon^2} + 4\left(\frac{X \ln \epsilon}{1 - \epsilon^2}\right)^2 + \mathcal{O}\left(\left[\frac{X \ln \epsilon}{1 - \epsilon^2}\right]^3\right). \qquad (3.16)$$

As displayed in Fig. 3.3, the corresponding convergence rates decrease only slightly as ϵ becomes smaller, which allows us to reach machine accuracy with moderate spectral resolution. We use this logarithmic coordinate mapping often to resolve steep gradients and can thus compute critical and limiting configurations very accurately.

4. In the last example we consider functions whose m-th derivative possesses a jump at some interior point $x_0 \in (0, 1)$, i.e. f is only C^{m-1}. It is well known that the so-called Gibbs phenomenon appears: the spectral approximation of the m-th derivative differs from its true value at x_0 by an amount independent of the spectral resolution order n (see Fig. 3.4). The spectral coefficients $c_k^{(n)}$ of f fall off as $k^{-(m+1)}$ (for $n \gg 1$) as reflected by the linear behaviour in Fig. 3.5, where the

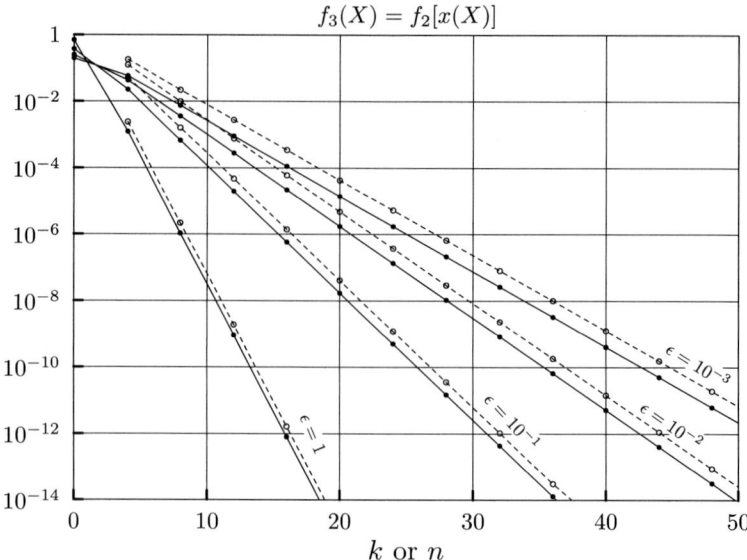

Fig. 3.3. Chebyshev coefficients $|c_k^{(100)}|$ (solid lines) and maximal residuals $R_{\max}^{(n)}$ (dashed lines) of the functions $f_3(X) = f_2[x(X)]$ for several values of ϵ.

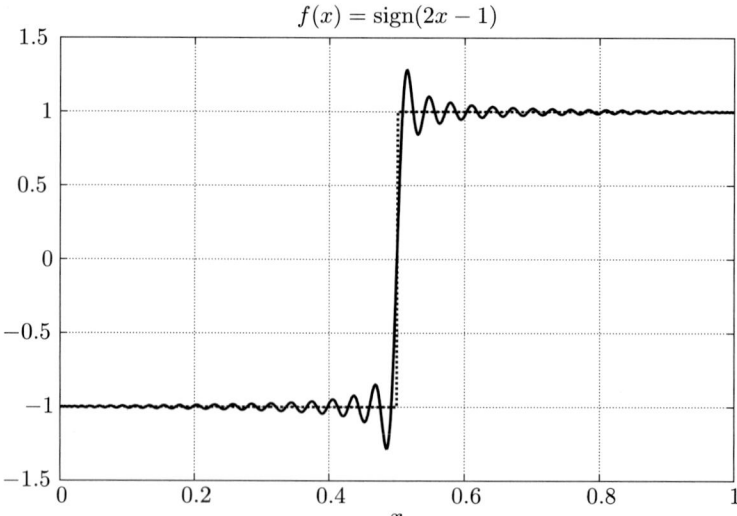

Fig. 3.4. Illustration of the Gibbs phenomenon. The spectral approximation of a discontinuous function differs from its true value at the location of the discontinuity by an amount independent of the spectral resolution order. The spectral approximation to 100th order (solid curve) is shown atop the function $f(x) = \text{sign}(2x - 1)$ (dashed curve).

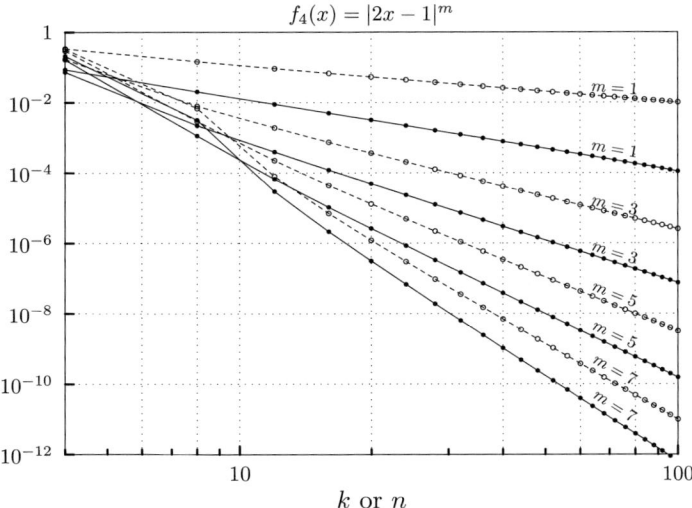

Fig. 3.5. Chebyshev coefficients $|c_k^{(200)}|$ (solid lines) and maximal residuals $R_{\max}^{(n)}$ (dashed lines) of the functions f_4 for several values of m.

resolution order n is plotted on a logarithmic scale. Note that the residuals $R_{\max}^{(n)}$ fall off at the smaller rate n^{-m}.

3.1.2 Solving differential equations with pseudo-spectral methods

Given the spectral coefficients $c_k^{(n)}$ of an analytic function f, the corresponding Chebyshev coefficients $c'^{(n)}_k$ of the derivative $f' = df/dx$ can be computed using

$$c'^{(n)}_{k-1} = \left(1 - \frac{1}{2}\delta_{k,1}\right)\left(c'^{(n)}_{k+1} + 4kc_k^{(n)}\right), \quad k = n-1, n-2, \ldots, 1, \qquad (3.17)$$

where $c'^{(n)}_{n-1} = c'^{(n)}_n = 0$ because of the fact that the coefficients $c_k^{(n)}$ do not contain information about the $(n-1)$-st and n-th coefficients of f'. Consequently, performing numerical derivatives by calculating the above coefficients leads to a drop in the accuracy available, as can be seen in Fig. 3.6. Nevertheless, the second derivative can still be calculated to high precision, which makes the pseudo-spectral approximation scheme applicable to the solution of second order elliptic boundary value problems.

In what follows, we describe the pseudo-spectral collocation point method, which forms the basis of the numerical techniques used in this book. Specifically, we consider second order, ordinary differential equations of the form

$$F(f, f', f''; x) = 0, \qquad (3.18)$$

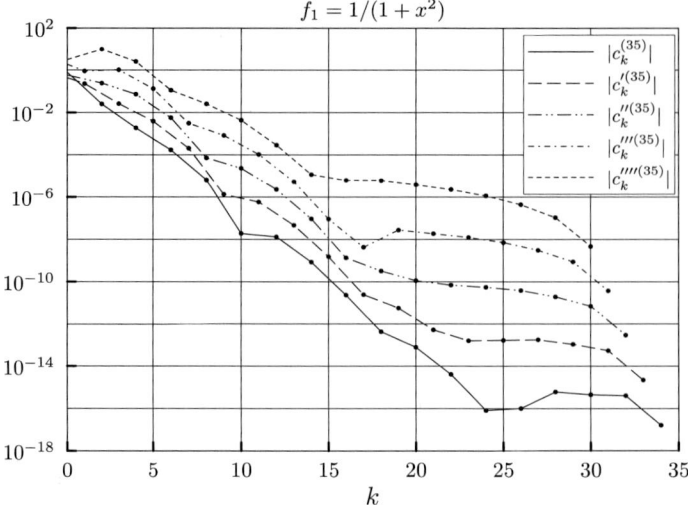

Fig. 3.6. Numerical calculation of the Chebyshev coefficients of f_1 and its derivatives f_1', f_1'', f_1''' and f_1'''' with 16 digits of numerical precision. The coefficients of the derivatives are computed using Equation (3.17) successively. For each additional derivative, approximately two digits of accuracy are lost due to the numerical evaluation.

with boundary conditions

$$F_0(f(0), f'(0)) = 0, \qquad F_1(f(1), f'(1)) = 0. \qquad (3.19)$$

The pseudo-spectral collocation point method approximates the values

$$f_j^{(n)} = f\left(x_j^{(n)}\right), \qquad (3.20)$$

which are combined to form the vector

$$\mathbf{f}^{(n)} = \begin{pmatrix} f_0^{(n)} \\ \vdots \\ f_{n-1}^{(n)} \end{pmatrix}. \qquad (3.21)$$

From any such vector $\mathbf{f}^{(n)}$, the spectral coefficients $c_k^{(n)}$, $c_k'^{(n)}$ and $c_k''^{(n)}$ of f, f' and f'' can be computed through Equations (3.8) and (3.17). The corresponding vectors $\mathbf{f}'^{(n)}$ and $\mathbf{f}''^{(n)}$, containing the values of f' and f'' at the collocation points $x_j^{(n)}$, are then approximated by the spectral representation (3.4), where the residual terms are neglected. With $\mathbf{f}^{(n)}, \mathbf{f}'^{(n)}$ and $\mathbf{f}''^{(n)}$, the given differential equation (3.18) and boundary conditions (3.19) yield the vector

$$\mathbf{F}^{(n)} = \begin{pmatrix} F_0^{(n)} \\ \vdots \\ F_{n-1}^{(n)} \end{pmatrix} \qquad (3.22)$$

with

$$F_j^{(n)} = \begin{cases} F_0\left(f_0^{(n)}, f_0^{\prime\,(n)}\right) & j = 0 \\ F\left(f_j^{(n)}, f_j^{\prime\,(n)}, f_j^{\prime\prime\,(n)}; x_j^{(n)}\right) & 0 < j < n-1 \\ F_1\left(f_{n-1}^{(n)}, f_{n-1}^{\prime\,(n)}\right) & j = n-1. \end{cases} \qquad (3.23)$$

The solution $\mathbf{f}^{(n)}$ of the discrete algebraic system

$$\mathbf{F^{(n)}}(\mathbf{f^{(n)}}) = 0 \qquad (3.24)$$

describes the spectral approximation of the solution f to the problem (3.18) and (3.19). We find the vector $\mathbf{f}^{(n)}$ using a Newton–Raphson scheme,

$$\mathbf{f}^{(n)} = \lim_{m \to \infty} \mathbf{f}_m^{(n)}, \quad \mathbf{f}_{m+1}^{(n)} = \mathbf{f}_m^{(n)} - \left[\mathbf{J}^{(n)}(\mathbf{f}_m^{(n)})\right]^{-1} \mathbf{F}^{(n)}\left(\mathbf{f}_m^{(n)}\right), \qquad (3.25)$$

where the Jacobian matrix is given by

$$\mathbf{J}^{(n)} = \frac{\partial \mathbf{F}^{(n)}}{\partial \mathbf{f}^{(n)}}. \qquad (3.26)$$

Note that for the convergence of the scheme, a 'good' initial guess $\mathbf{f}_0^{(n)}$ is necessary, which we provide through a known nearby function.

In order to illustrate the method, we consider three examples of linear equations:

$$(1+x^2)^2 f'' + 3x(1+x^2)f' + 2f = 0 \quad \text{with} \quad f(0) = 1, \quad f(1) = \frac{1}{2}, \qquad (3.27)$$

$$x(x-1)f'' + (2x-1)f' - f = -\frac{x(x+5)}{(x+1)^3} \qquad (3.28)$$

and

$$x^2(1-x)^2 f'' + 4x(1-x)^2 f' + 2(1-2x)f = x^2 - \frac{2}{9}(1+x). \qquad (3.29)$$

The differential equation (3.27) is non-singular, and the solution reads $f = f_1(x)$, with f_1 as defined in (3.11).

On the contrary, Equation (3.28) is weakly singular at the points $x = 0$ and $x = 1$. The solution is given by

$$f(x) = \frac{1}{1+x} + c_1 P_\nu(2x-1) + c_2 Q_\nu(2x-1), \quad \nu = \frac{1}{2}\left(-1+\sqrt{5}\right), \quad (3.30)$$

where P_ν and Q_ν are Legendre functions of the first and second kind. This expression is regular for $0 \leq x \leq 1$ only if $c_1 = c_2 = 0$, in which case $f = f_2$ with f_2 as defined in (3.13) for $\epsilon = 1$. Consequently, it is not possible to impose arbitrary boundary values, but, for a regular solution, Equation (3.28) implies that

$$f'(0) + f(0) = 0, \quad f'(1) - f(1) = -\frac{3}{4}. \quad (3.31)$$

Weakly singular equations such as (3.28) appear quite commonly in the subsequent sections, which is due to the specific coordinate mappings being used. It is therefore essential to ensure that the pseudo-spectral method provides the corresponding regular solutions from the information given by the equations.

Finally, Equation (3.29) describes a situation in which no analytic solution on $[0, 1]$ exists. For a bounded solution with $x^2(1-x)^2 f''$ as well as $x(1-x)^2 f'$ tending to zero as $x \to 0$ or $x \to 1$, the differential equation implies the boundary conditions

$$f(0) = -\frac{1}{9}, \quad f(1) = -\frac{5}{18}. \quad (3.32)$$

The corresponding solution reads

$$f_{C^1}(x) = \frac{1}{18x^2}\left[x(6-11x) + 6(1-x)^2 \ln|1-x|\right]. \quad (3.33)$$

This function is analytic at $x = 0$ but only C^1 at $x = 1$, where a logarithmic singularity is present. As illustrated in Appendix A1.1, we have to deal with functions of this kind if we encounter the mass-shedding limit of rotating stars, see also Fig. 3.11 below.

We have chosen the examples (3.27), (3.28, 3.31) and (3.29, 3.32) with the solutions f_1, f_2 and f_{C^1} in order to provide a comparison of the numerical solution with the direct spectral approximation described in the previous section. As can be seen from Figs 3.7–3.9, the fall-off in the maximal residuals $R_{\max}^{(n)}$ resulting from (i) the numerical integration of the differential equations to the approximation order n, and (ii) the analytically known Chebyshev coefficients, is very similar. In examples (3.27) and (3.28, 3.31) the numerical results saturate when an accuracy of about 14 digits is reached, which is due to the internal numerical precision of 16 digits being used.

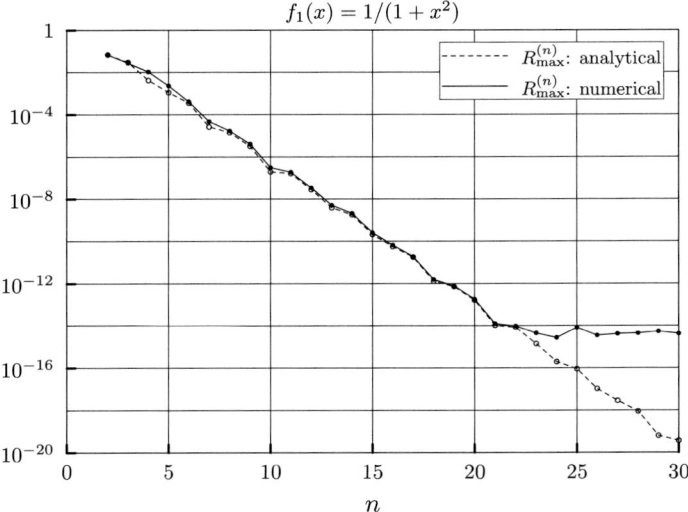

Fig. 3.7. Maximal residuals $R_{\max}^{(n)}$ resulting from (i) the numerical integration of (3.27) to the approximation order n, and (ii) the analytically known Chebyshev coefficients.

Note that in Fig. 3.9, an algebraic fall-off with a slope of

$$m_{\text{algebr}}(n) = \frac{\ln R_{\max}^{(n)}}{\ln n} \approx -4$$

results. This seems to be in contradiction with the statement made at the end of Subsection 3.1.1, claiming that for a C^{m-1}-function, a slope $m_{\text{algebr}}(n) = -m$ as $n \to \infty$ is to be expected. The better convergence rate exhibited for the example (3.29, 3.32) is due to the fact that, contrary to the examples f_4 in Fig. 3.5, the singular behaviour is located at the boundary of the interval being considered.

We conclude this section by abandoning toy functions and providing examples of convergence for two numerically generated figures of equilibrium that are amongst those to be discussed in Section 3.3. Each is governed by the equation of state for homogeneous matter, $\epsilon = $ constant. The first was generated by prescribing the central pressure $p_c/\epsilon = 1$ and radius ratio[1] $A = 0.7$, a configuration that is discussed in Nozawa *et al.* (1998) and Ansorg *et al.* (2003a). The second is described by prescribing the same central pressure $p_c/\epsilon = 1$ and requiring that it rotate at the mass-shedding limit. The measure we choose for the accuracy of the

---

[1] See (3.46) for a definition of the radius ratio A.

126   *Numerical treatment of the general case*

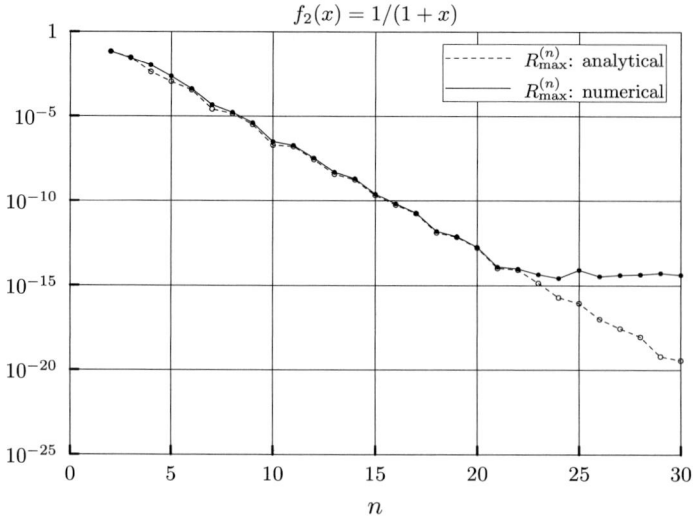

Fig. 3.8. Maximal residuals $R_{\max}^{(n)}$ resulting from (i) the numerical integration of (3.28, 3.31) to the approximation order n, and (ii) the analytically known Chebyshev coefficients.

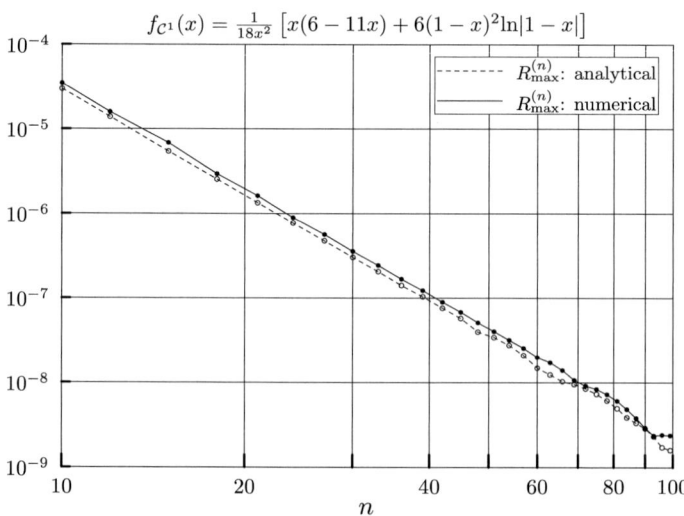

Fig. 3.9. Maximal residuals $R_{\max}^{(n)}$ resulting from (i) the numerical integration of (3.29, 3.32) to the approximation order n, and (ii) the analytically known Chebyshev coefficients.

solutions is

$$R_\phi^{(n)} = \max_{(s,t)\in\{(s_i^{(\mathrm{ref})}, t_j^{(\mathrm{ref})})\}} \left\{ \left| \frac{\phi_k^{(\mathrm{ref})}(s,t) - \phi_k^{(n)}(s,t)}{\phi_k^{(\mathrm{ref})}\,{}_{\mathrm{central}}} \right| \right\}, \qquad (3.34)$$

where each ϕ_k refers to one of the four metric functions, $s_i^{(n)}$ refers to the spectral gridpoints of Equation (3.7) and the superscript 'ref' refers to a reference solution. To put it in words, after a reference solution is chosen, the value of each metric function, normalized with respect to its central value, is compared to the value at

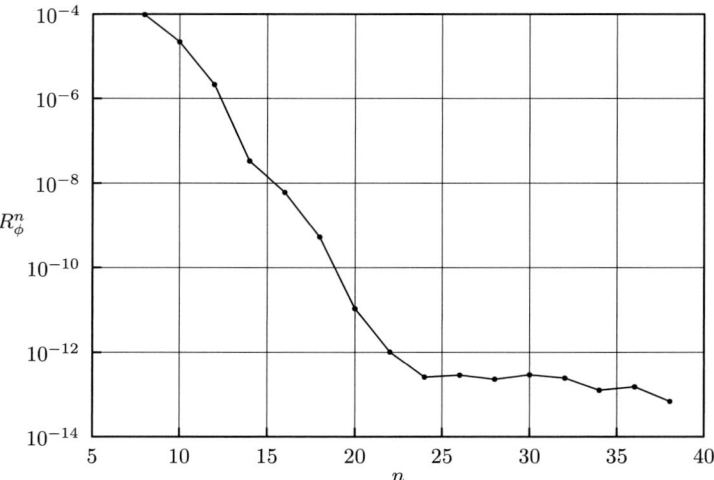

Fig. 3.10. The convergence is shown for the numerical solution describing a homogeneous star with central pressure $p_c/\epsilon = 1$ and radius ratio $A = 0.7$.

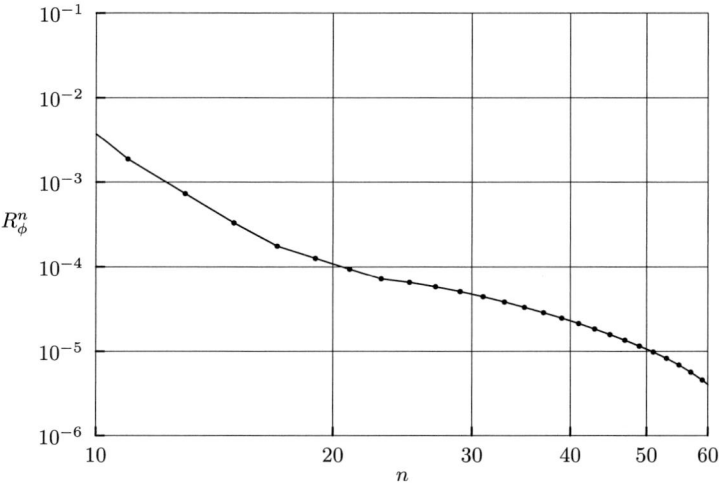

Fig. 3.11. The convergence is shown for the numerical solution describing a homogeneous star with central pressure $p_c/\epsilon = 1$ and rotating at the mass-shedding limit.

the same (s, t) point for a solution of a different order. After scanning through every gridpoint in each domain, the maximum absolute value is chosen as the measure.

In Fig. 3.10 one can see that the convergence for the first of the two configurations is similar to Figs 3.7 and 3.8. The metric functions being approximated are analytic and the resulting solution converges to a level determined by the internal numerical precision. As discussed in Appendix A1.1, the second configuration, which rotates at the mass-shedding limit, contains functions that are only C^1. The singular behaviour is to be found along the equator, which is located at the boundary of the relevant interval in s and t. It is then not surprising that in Fig. 3.11, showing the convergence of this numerical solution, we find behaviour somewhat reminiscent of Fig. 3.7 although not as linear. In the region with $n > 25$, the convergence curve can be approximated by a line with the slope

$$m_{\text{algebr}}(n) = \frac{\ln R_\phi^{(n)}}{\ln n} \approx -3.$$

3.2 Coordinate mappings

As demonstrated in Subsection 3.1.1, exponential convergence for the spectral expansions can only be obtained if the functions are analytic within the domain considered. At the surface of the fluid configuration, higher derivatives of the metric potentials are discontinuous, leading to Gibbs phenomena when spectrally expanded. Therefore, we choose to divide up the entire spatial region into several subdomains, where one of them coincides exactly with the interior of the fluid. In this manner we avoid Gibbs phenomena because the location of the discontinuities is at the subdomain's boundary and the functions can be expanded in Taylor series converging on either side of it.[2]

Each of these subdomains is characterized by a mapping,

$$\varrho^2 = \varrho_k^2(s, t), \qquad \zeta^2 = \zeta_k^2(s, t), \qquad (s, t) \in [0, 1]^2$$

where k labels the subdomains ($k = 0, \ldots, n_{\text{dom}} - 1$). Since for a free boundary value problem, the surface shape of the fluid body is not known a priori, a one-dimensional function G,

$$G : [0, 1] \to \mathbb{R},$$

is involved that describes the geometrical boundary.

---

[2] Note that for equations of state, in which the pressure $p(h)$ and/or energy density $\epsilon(h)$ are not analytic with respect to the enthalpy h at $h = h(0)$, the metric functions are also not analytic, and only an algebraic convergence rate results, see Fig. 3.26.

3.2.1 Mappings for spheroidal configurations

For spheroidal configurations, we introduce two subdomains, i.e.

$$n_{\text{dom}} = 2,$$

with the domain $k = 0$ covering the entire vacuum space exterior to the fluid body (compactifying spatial infinity), and the domain $k = 1$ covering the fluid's interior.[3] An example of the coordinates used can be found in Fig. 3.12 and the particular functions are given below.

- **Exterior subdomain ($k = 0$):**

$$\varrho_0^2(s,t) = t\left[r_e^2 - r_p^2 + \xi^2(s)\right],$$
$$\zeta_0^2(s,t) = (1-t)\left[\xi^2(s) - r_p^2\right] + G(t) - r_e^2 t \quad (3.35a)$$

with

$$\xi(s) = r_p + r_e \frac{1 - \sigma(s)}{\sigma(s)},$$
$$\sigma(s) = 1 - \frac{\sinh[(1-s)\ln \varepsilon_s]}{\sinh(\ln \varepsilon_s)}, \quad (3.35b)$$

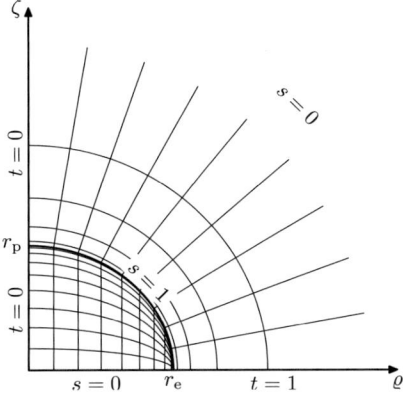

Fig. 3.12. Coordinate lines of constant s and t are shown for an example of a spheroidal configuration, where the surface of the configuration, $s = 1$, is indicated by a thicker line. The coordinates depicted here are particularly well suited for working with very flat configurations.

---

[3] With the coordinate mappings in question we are able to compute stars with extremely large central pressure $p_c \approx 10^5 \epsilon_c$. However, for stars with even larger (including infinite) central pressure we refer the reader to the potential representations and coordinates used in Ansorg et al. (2003a).

$$G(t=0) = r_{\rm p}^2,$$
$$G(t=1) = r_{\rm e}^2. \tag{3.35c}$$

The boundaries of the exterior domain are described by

$s = 0$: spatial infinity, $\sqrt{\varrho^2 + \zeta^2} \to \infty$

$s = 1$: surface of the fluid, given by:

$$\left\{ (\varrho, \zeta) = \left(r_{\rm e}\sqrt{t}, \sqrt{G(t) - r_{\rm e}^2 t} \right), t \in [0, 1] \right\} \tag{3.36}$$

$t = 0$: rotation axis, $\varrho = 0$

$t = 1$: equatorial plane, $\zeta = 0$.

- **Interior subdomain ($k = 1$):**

$$\varrho_1^2(s,t) = r_{\rm e}^2 t,$$
$$\zeta_1^2(s,t) = s \left[G(t) - r_{\rm e}^2 t \right]. \tag{3.37}$$

The boundaries of the interior domain are described by:

$s = 0$: equatorial plane, $\zeta = 0$

$s = 1$: surface of the fluid, as in (3.36)

$t = 0$: rotation axis, $\varrho = 0$ \hfill (3.38)

$t = 1$: equator, $\varrho = r_{\rm e}, \zeta = 0$.

The mapping of the exterior subdomain is chosen to resemble oblate spheroidal coordinates in which the entire class of Maclaurin spheroids exhibits a rapid spectral convergence rate. For highly flattened relativistic stars we find, however, that the spectral convergence rate can be improved considerably by refining the exterior spectral mesh in the vicinity of the fluid's surface. We do so by rescaling the coordinate s in the same manner as in Example 3 of Subsection 3.1.1 and adjusting the free parameter ε_s.

3.2.2 Mappings for toroidal configurations

If one deals with toroidal configurations, it is natural to make use of toroidal coordinates. In particular, we describe the vacuum region exterior to the fluid ring

3.2 Coordinate mappings

by means of the coordinates $(\tilde{x}, \tilde{y})$ with

$$\varrho^2 = \varrho_m^2 \frac{\tilde{y}(\tilde{y}+1)}{(\tilde{x}+\tilde{y})^2},$$
$$\zeta^2 = \varrho_m^2 \frac{\tilde{x}(1-\tilde{x})}{(\tilde{x}+\tilde{y})^2}, \tag{3.39}$$

from which it follows that

$$r^2 = \varrho^2 + \zeta^2 = \varrho_m^2 \frac{1+\tilde{y}-\tilde{x}}{\tilde{x}+\tilde{y}}. \tag{3.40}$$

Here the parameter ϱ_m must be within the interval (ϱ_i, ϱ_o) where $\varrho_{i/o}$ are the inner and outer coordinate radii of the ring respectively. We choose

$$\varrho_m = \frac{1}{2}(\varrho_i + \varrho_o). \tag{3.41}$$

The inverse of transformation (3.39) reads

$$\tilde{x} = \frac{1}{2}\left(1 + \frac{\varrho_m^2 - \varrho^2 - \zeta^2}{\sqrt{(\varrho_m^2 - \varrho^2)^2 + 2\zeta^2(\varrho^2 + \varrho_m^2) + \zeta^4}}\right),$$
$$\tilde{y} = \frac{1}{2}\left(-1 + \frac{\varrho_m^2 + \varrho^2 + \zeta^2}{\sqrt{(\varrho_m^2 - \varrho^2)^2 + 2\zeta^2(\varrho^2 + \varrho_m^2) + \zeta^4}}\right). \tag{3.42}$$

The coordinates $(\tilde{x}, \tilde{y})$ are bounded,

$$0 \leq \tilde{x} \leq 1, \qquad 0 \leq \tilde{y} \leq \tilde{y}_b(\tilde{x}), \tag{3.43}$$

where

- $\tilde{x} = 0$ describes the portion of the equatorial plane beyond the exterior edge of the ring, i.e. $\varrho \geq \varrho_o, \zeta = 0$,
- $\tilde{x} = 1$ describes the portion of the equatorial plane between the rotation axis and interior edge of the ring, i.e. $0 \leq \varrho \leq \varrho_i, \zeta = 0$,
- $\tilde{y} = 0$ describes the rotation axis, i.e. $\varrho = 0$, and
- $\tilde{y} = \tilde{y}_b(\tilde{x})$ describes the (unknown) surface shape of the ring.

We divide this exterior domain into four subdomains. The interior of the ring contributes a further subdomain such that

$$n_{\text{dom}} = 5.$$

Note that the following coordinate mappings, when discussed in neighbouring domains, give the same description of the mutual border. Representative plots displaying the partition of the spatial domain as well as the corresponding coordinate lines of constant s and t values are provided in Fig 3.13.

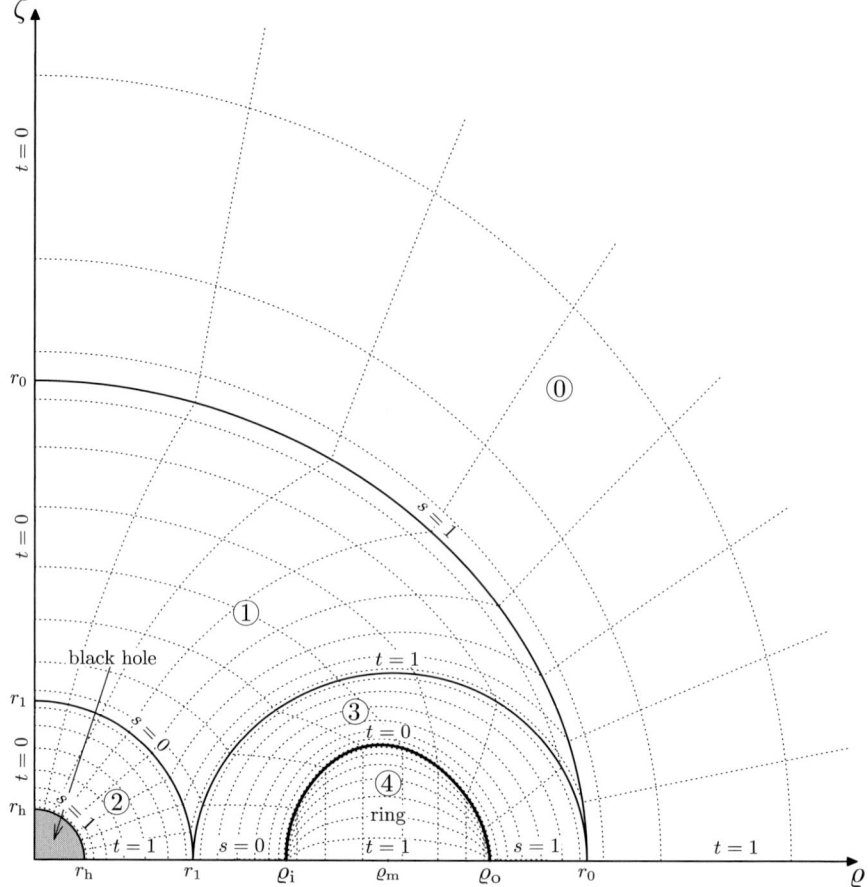

Fig. 3.13. Coordinate lines of constant s and t are shown for an example of a toroidal configuration surrounding a central black hole. The surface of the ring is indicated by a thicker line (after Ansorg and Petroff 2005).

- **Exterior subdomains ($k = 0, 1, 2, 3$):**
  - **subdomain $k = 0$**
    This domain compactifies the exterior of a spherical shell with prescribed radius $r_0 > \varrho_0$. Constant s-values correspond to constant r-values:
    $$r = \frac{r_0}{s},$$
    $$\tilde{x}_0 = \frac{\varrho_m^2 s^2 (1-t)}{r_0^2 + \varrho_m^2 s^2 (1-2t)},$$
    $$\tilde{y}_0 = \frac{\varrho_m^2 s^2 t}{r_0^2 + \varrho_m^2 s^2 (1-2t)}.$$

3.2 Coordinate mappings

– **subdomain $k = 1$**

The boundaries of this domain are two spherical shells (with radii r_0 and $r_1 < \varrho_i$), a section of the rotation axis ($\zeta \in [r_1, r_0]$) and a toroidal curve around the ring, intersecting the equatorial plane at the radii r_1 and r_0. As with subdomain 0, constant s-values correspond to constant r-values:

$$r = r_1 \sigma_1 = r_1 \left(\frac{r_0}{r_1}\right)^s,$$

$$\tilde{x}_1 = \frac{\varrho_m^2 (r_0^2 - r_1^2) - r_1^2 t[\sigma_1^2(\varrho_m^2 - r_1^2) + (r_0^2 - \varrho_m^2)]}{(r_0^2 - r_1^2)(\varrho_m^2 + r_1^2 \sigma_1^2) - 2r_1^2 t[\sigma_1^2(\varrho_m^2 - r_1^2) + (r_0^2 - \varrho_m^2)]},$$

$$\tilde{y}_1 = \frac{r_1^2 t[\sigma_1^2(\varrho_m^2 - r_1^2) + (r_0^2 - \varrho_m^2)]}{(r_0^2 - r_1^2)(\varrho_m^2 + r_1^2 \sigma_1^2) - 2r_1^2 t[\sigma_1^2(\varrho_m^2 - r_1^2) + (r_0^2 - \varrho_m^2)]}.$$

– **subdomain $k = 2$**

This domain is a spherical shell around the central object which can either be a black hole (with the horizon chosen to be a sphere in the coordinates (ϱ, ζ)) or an infinitely flattened disc of dust. In the case of a single toroidal configuration without a central object, we may describe the central vacuum region by a disc of dust with vanishing surface mass-density, i.e. we may take the same coordinates as for configurations with central discs.

For a central black hole (with horizon coordinate radius $r_h < r_1$), we take a shell with inner radius r_h and outer radius r_1. Again, constant s-values correspond to constant r-values:

$$r = r_1 \sigma_2 = r_1 \left(\frac{r_h}{r_1}\right)^{1-s},$$

$$\tilde{x}_2 = \frac{\varrho_m^2 \sigma_2^2 - r_h^2 t}{\varrho_m^2 \sigma_2^2 + r_h^2 (1 - 2t)},$$

$$\tilde{y}_2 = \frac{r_h^2 t}{\varrho_m^2 \sigma_2^2 + r_h^2 (1 - 2t)}.$$

For a central disc (with coordinate radius $\varrho_d < r_1$) we take a transformation that interpolates between oblate spheroidal coordinates in the vicinity of the disc and spherical coordinates in the vicinity of the shell $r = r_1$:

$$\varrho_2^2 = \left[\varrho_d^2 + (s-1)^2 \xi^2(t)\right] \tau(t),$$

$$\zeta_2^2 = (s-1)^2 \xi^2(t) [1 - \tau(t)]$$

with

$$\xi^2(t) = \frac{1}{2}\left[r_1^2 - \varrho_d^2 + \sqrt{(r_1^2 + \varrho_d^2)^2 - 4tr_1^2\varrho_d^2\left(1 + \frac{r_1^2}{\varrho_m^2}(1-t)\right)}\right],$$

$$\tau(t) = \frac{r_1^2 t\left[\varrho_m^2 + r_1^2(1-t)\right]}{\varrho_m^2\left[\varrho_d^2 + \xi^2(t)\right]}.$$

- **subdomain $k = 3$**
  This domain is a toroidal shell around the ring. The outer toroidal boundary intersects the equatorial plane at the radii r_1 and r_0, and the inner one at ϱ_i and ϱ_o:

$$\tilde{x}_3 = t\tilde{x}_1(s, t = 1) + (1 - t)\tilde{x}_4(s, t = 0),$$

$$\tilde{y}_3 = \tilde{y}_4(s, t = 0)\left(\frac{\tilde{y}_1(s, t = 1)}{\tilde{y}_4(s, t = 0)}\right)^t.$$

The coordinates $\tilde{x}_4$ and $\tilde{y}_4$ follow from ϱ_4^2 and ζ_4^2 defined below through transformation (3.42).

- **Interior subdomain $k = 4$**
  This domain covers the interior of the ring. Its surface is obtained for $t = 0$ whereas $t = 1$ yields a section of the equatorial plane that is inside the ring:

$$\varrho_4^2 = \varrho_i^2(\varrho_o/\varrho_i)^{2s},$$

$$\zeta_4^2 = (1 - t)[G(s) - \varrho_4^2].$$

In this transformation, the function G appears explicitly, which, as with spheroidal configurations, describes the unknown surface shape of the perfect fluid body. From G, the alternative representation $\tilde{y}_b$, introduced in (3.43), is given implicitly by the relation

$$\tilde{y}_b[\tilde{x}_4(s, t = 0)] = \tilde{y}_4(s, t = 0).$$

As the inner and outer equatorial edges of the ring correspond to $s = 0$ and $s = 1$ respectively, the surface function G is subject to the conditions

$$G(s = 0) = \varrho_i^2, \qquad (3.44)$$

$$G(s = 1) = \varrho_o^2. \qquad (3.45)$$

The function G (and hence the coordinate radii ϱ_i and ϱ_o) is not known, but is determined through the numerical solution procedure. In the case of a toroidal configuration with a central object, this also applies to the coordinate radii r_h or ϱ_d. On the contrary, the parameters r_0 and r_1 can be prescribed freely for the calculation.

In order to get good accuracy even when using low resolution, these parameters should be chosen within some optimal interval.

It will be useful in later sections to be able to refer to a radius ratio, A, for spheroidal as well as toroidal configurations. We thus define

$$A := \begin{cases} \frac{r_p}{r_e}: & \text{for spheroidal configurations} \\ -\frac{\varrho_i}{\varrho_o}: & \text{for toroidal configurations} \end{cases} \quad (3.46)$$

to be the ratio of polar to equatorial coordinate radius for spheroidal configurations and the negative of the inner to outer coordinate radius for toroidal ones. The radius ratio runs from -1 in the thin ring limit to 1 in the spherical limit, and $A = 0$ marks the transition point from one topology to the other.

3.2.3 The solution vector f of the Newton–Raphson scheme

The numerical scheme to treat the free boundary value problems for axisymmetric, stationary equilibrium configurations generalizes the pseudo-spectral collocation point method introduced in Subsection 3.1.2 for ordinary differential equations. All functions U^κ ($\kappa = 0, 1, \ldots, n_{\text{eq}} - 1$) to be determined by the free boundary value problem are considered at specific gridpoints $(s_{k,i}; t_{k,j})$ in the subdomains $k = 0, 1, \ldots, n_{\text{dom}}$. These gridpoints are given through Equation (3.7), that is:

$$s_{k,i} = \sin^2\left(\frac{\pi i}{2(n_k^{(s)} - 1)}\right) \;;\; i = 0, 1, \ldots, n_k^{(s)} - 1;$$
$$t_{k,j} = \sin^2\left(\frac{\pi j}{2(n_k^{(t)} - 1)}\right) \;;\; j = 0, 1, \ldots, n_k^{(t)} - 1. \quad (3.47)$$

The numbers $n_k^{(s)}$ and $n_k^{(t)}$ of gridpoints in domain k with respect to the s and t directions (i.e. the spectral expansion orders) may assume different values in the various subdomains, but at common boundaries the numbers have to coincide, i.e. for spheroidal configurations we have

$$n_0^{(t)} = n_1^{(t)},$$

and for toroidal ones:

$$n_0^{(t)} = n_1^{(t)} = n_2^{(t)},$$
$$n_1^{(s)} = n_3^{(s)} = n_4^{(s)}.$$

As in Subsection 3.1.2, we collect all function values

$$U_{k,ij}^\kappa = U^\kappa(s_{k,i}, t_{k,j}) \quad (3.48)$$

as well as the values of the unknown surface function G,

$$G_j = G(t_{1,j}) \quad \text{for spheroidal configurations and}$$
$$G_i = G(s_{4,i}) \quad \text{for toroidal configurations,}$$

in order to build up a vector $\mathbf{f}$. In addition, this vector is filled with a number n_par of physical parameters that characterize the configuration. For a given equation of state, $n_\text{par} = 2$ for one-body configurations, and we choose them to be V_0 and Ω, cf. Section 1.4. If a central object is present, then there are another two parameters (i.e. $n_\text{par} = 4$) for which we take the angular velocity and coordinate radius of the central object, that is $(\Omega_\text{h}, r_\text{h})$ for a black hole and $(\Omega_\text{d}, \varrho_\text{d})$ for a disc. Whereas the number of parameters n_par is fixed, it is sometimes possible to find more than one solution to a given set of parameters. For example, the crossing of the dashed and the solid lines in the inset of Fig. 3.17 below demonstrates that (at least) two configurations exist for that $\Omega/\sqrt{\epsilon}$-A pair.

The collection of elliptic equations valid in the subdomains, the transition conditions at common domain boundaries, the vanishing pressure boundary condition at the fluid's surface and certain parameter relations that one wishes to fulfil, yield a discrete nonlinear system of the form (3.24) where n stands for the collection of all $n_k^{(s)}$ and $n_k^{(t)}$,

$$n = \{(n_k^{(s)}, n_k^{(t)}); \quad k = 0, 1, \ldots, n_\text{dom} - 1\}.$$

The dimension of this system is given by

$$n_\text{total} = n_\text{eq} \sum_{k=0}^{n_\text{dom}-1} n_k^{(s)} n_k^{(t)} + n_\text{G} + n_\text{par}, \tag{3.49}$$

with $n_\text{G} = n_1^{(t)}$ for spheroidal and $n_\text{G} = n_4^{(s)}$ for toroidal configurations. In particular, the transition conditions require the U^κ to be continuous and to possess continuous normal derivatives. At domain boundaries that correspond to portions of the rotation axis or the equatorial plane, we require regularity conditions, which follow from the elliptic equations when specialized to this boundary. Via the integrated relativistic Euler equation (3.3), the vanishing pressure boundary condition restricts the potentials at the fluid's surface. It adds n_G equations to the system. Finally, we may include specific parameter relations that we wish to be satisfied. For example, we could simply prescribe certain values for the physical parameters contained in $\mathbf{f}$. However, we may instead wish to prescribe other parameters, say gravitational mass and angular momentum of the objects. For this reason we include the n_par physical parameters into the vector $\mathbf{f}$ and add the appropriate parameter relations to the system.

3.3 Equilibrium configurations of homogeneous fluids

In the previous section, a numerical method was presented for calculating figures of equilibrium. In this section, such methods will be applied to single homogeneous bodies, i.e. those described by the equation of state $\epsilon =$ constant, where ϵ is the energy-density. Newtonian homogeneous figures of equilibrium have played an invaluable role in the history of gravitational theory because of their simplicity, tractability and multifariousness. A number of the key contributors to the field were listed in the preface, and important books combining both old and new material were written by Lichtenstein (1933) and Chandrasekhar (1969). The latter went on to consider relativistic homogeneous figures of equilibrium from various vantage points.

The importance of this equation of state for Newtonian as well as Einsteinian theory is underscored by the number of analytic solutions that can be found using it. The Maclaurin spheroids, including their spherical and disc limits, have been presented here. Schwarzschild spheres and the rigidly rotating relativistic disc of dust, which are attached to these end points in the Newtonian limit, were also treated at length. It was found that the exterior extreme Kerr metric results as the extreme relativistic limit for the disc of dust. The question that arises is how these motley solutions fit into the larger picture of solutions to Einstein's equations for homogeneous matter.

To answer this question, we employ a combination of analytic and numerical techniques. Analytic results will provide us with our first 'good initial guess' for the Newton–Raphson scheme, cf. page 123. Furthermore they will allow us to locate singular configurations that are important for distinguishing between distinct regions of solution space. By beginning at a known analytic solution, say the Schwarzschild spheres, and tracing out connected boundaries in this solution space, it will be possible to divide it into a countably infinite number of distinct classes. A schematic portrayal of the picture that emerges is given in Fig. 3.14, which will serve us throughout this section as a roadmap.

Our task now is to proceed to explain how this class structure was ascertained and what its basic properties are.

3.3.1 Bifurcation points

We have already encountered Newtonian homogeneous figures of equilibrium in Section 2.1, namely the Maclaurin spheroids. The secular stability of these and other figures of equilibrium was studied by Poincaré (1885), who determined the change in energy associated with a change in the shape of the surface up to second order. He focused primarily on the second harmonic, non-axially symmetric

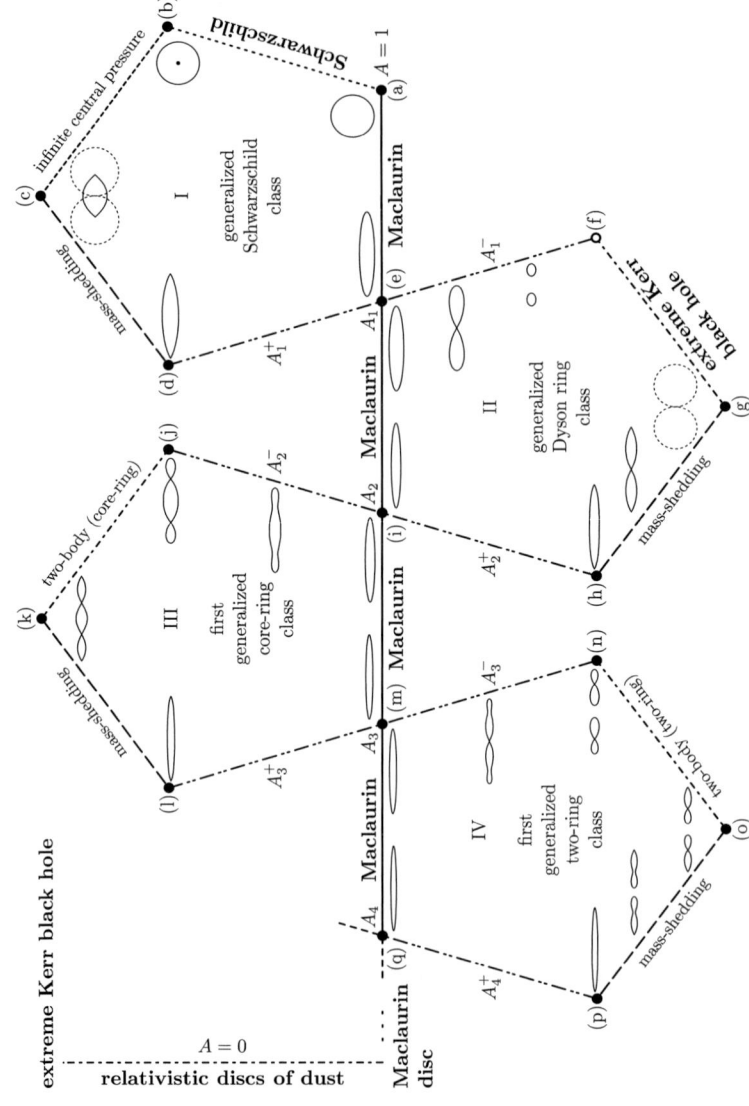

Fig. 3.14. A schematic portrayal of the classes and their boundaries. The analytically known solutions are written in larger bold-faced letters (after Ansorg *et al.* 2004).

3.3 Equilibrium configurations of homogeneous fluids

Table 3.1. *Numerical values for the first four solutions to Equation (3.50). In addition to the value of ξ_0, the corresponding values of the eccentricity ϵ and of the ratio of the minor to major axis of the ellipse are given.*

l	ξ_0	ϵ	$c/a = A_{l-1}$
2	0.17383011	0.98522554	0.17126187
3	0.11230482	0.99375285	0.11160323
4	0.08303471	0.99657034	0.08274493
5	0.06588682	0.99783651	0.06574427

perturbation. On the other hand, Chandrasekhar (1967, 1968) studied the fourth harmonic axially symmetric mode and determined the value for the eccentricity at the onset of axially symmetric instabilities. Bardeen (1971) derived formulae to determine all marginal, axisymmetric instabilities, which were generalized to include non-axisymmetric modes by Hachisu and Eriguchi (1984). Numerical results (Eriguchi and Hachisu 1982, Ansorg *et al.* 2003c) confirm that the solutions to these equations indeed mark bifurcation points along the Maclaurin sequence.

By considering an axisymmetric perturbation in the shape of the Maclaurin spheroids, it is possible to locate non-ellipsoidal Newtonian stars deviating only infinitesimally from an ellipsoid (see Ansorg *et al.* 2003c). Such configurations are found when the constant value of the elliptic coordinate describing the surface of the body is a solution of the equation

$$iP_{2l}(i\xi_0)Q_{2l}(i\xi_0) - \xi_0(1 - \xi_0 \operatorname{arccot}\xi_0) = 0; \quad l = 2, 3, 4, \ldots; \quad (3.50)$$

where $P_n(x)$ and $Q_n(x)$ are the Legendre polynomials and Legendre functions of the second kind. For a given value of l, there is exactly one solution to (3.50), which will be denoted by ξ_{2l}^*. The first few are listed in Table 3.1. These values for ξ_0 mark a countably infinite number of bifurcation points that have an accumulation point at $\xi_0 = 0$.

The nature of these bifurcation points within relativity is quite interesting. Chandrasekhar (1967) noticed that the first post-Newtonian expansion for the Maclaurin spheroids contains a singularity at the first bifurcation point associated with $l = 2$. Upon reexamining this work, Bardeen (1971) conjectured that there exists a singularity in the post-Newtonian expansion for every bifurcation point. This conjecture was verified by Petroff (2003), who showed that the singularity associated with the value l first appears at the $(l-1)$-st post-Newtonian order.

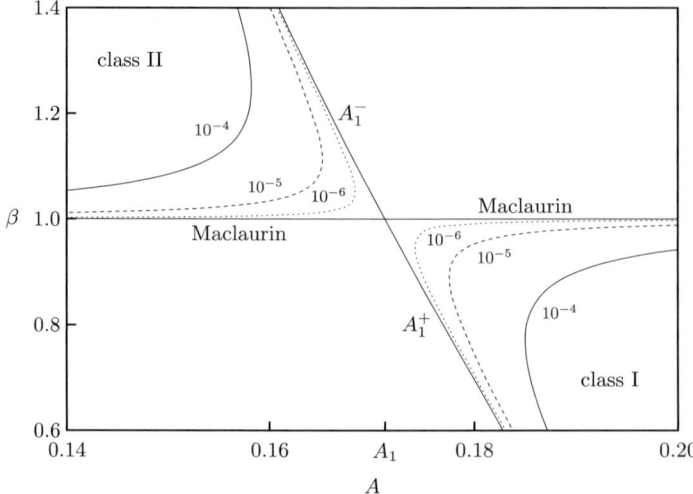

Fig. 3.15. Sequences of constant $M\sqrt{\epsilon}$ are plotted in the β-A parameter space in the vicinity of the first bifurcation point A_1. For a definition of the mass-shed parameter β, see (3.51), and for the radius ratio A, see (3.46) (after Ansorg et al. 2004).

If one assumes that the post-Newtonian expansion converges at least for some values of a relativistic expansion parameter, then these results imply that there exists no (relativistic) figure of equilibrium at these bifurcation points. This expectation is confirmed by numerical results. For example, Fig. 3.15 shows lines of constant normalized gravitational mass $M\sqrt{\epsilon}$ as it depends on the radius ratio A of (3.46) in the vicinity of the first bifurcation point. For any non-zero value of the mass, there is a region about the bifurcation point containing no solutions. What is more, we shall see that it is not possible to cross over from the solutions on one side of this point to those on the other without passing through the bifurcation point itself. This behaviour of the solution space turns out to hold for each of the bifurcation points and allows us to divide the solution space into classes. Each class contains a section of the Maclaurin sequence running from $\xi_0 = \xi_{2l}^*$ to $\xi_0 = \xi_{2l+2}^*$. The nature of the classes will become clearer in the next sections, when we examine the first few of them, comment on the remaining ones and finally sketch out a picture of the solution space.

3.3.2 Class I: the generalized Schwarzschild class

As can be seen in Fig. 3.14, the generalized Schwarzschild class is demarcated by five boundary sequences, one of which is the static limit. A detailed look at this class can be found in Schöbel and Ansorg (2003), which also provides some information

The Schwarzschild solution: Consider the interior and exterior Schwarzschild solution for a given energy density ϵ, as discussed in Section 2.2. The solution depends on one parameter, say the relative redshift z of zero angular momentum photons of Equation (1.28), which runs from 0 in the Newtonian limit to 2 in the limit of infinite central pressure ((a) to (b) in Fig. 3.14). The value for this redshift can be further increased if we abandon the static limit, allowing the star to rotate. Such a rotating star is described by two parameters. Let us choose to consider a sequence of stars with infinite central pressure and ask what happens when we increase the angular velocity Ω.

Infinite central pressure: Equation (2.30) shows that $e^{2\nu} \to 0$ at the star's centre as the central pressure $p_c \to \infty$. As soon as such a star is set in motion, an ergosphere forms. The ergosphere grows outward from the centre as Ω is increased and crosses the star's surface. Increasing Ω further, causes the shape of the surface at the equatorial plane to become more pointed until, finally, a cusp indicating a mass-shedding limit is reached ((b) to (c) in Fig. 3.14). This salient configuration deserves more attention, since many important physical properties, such as mass, attain their maximum values for this class precisely here, where the sequence of infinite central pressure meets the mass-shedding limit. As we can see in Table 3.2, the (polar) redshift can increase significantly relative to the non-rotating case. Furthermore, the maximal attainable mass, given by the Buchdahl limit for the Schwarzschild solution, i.e. Equation (2.37), increases by roughly 34% due to rotation. Figure 3.16 shows the coordinate shape of this star and its ergosphere.

The mass-shedding sequence: We can describe a sequence of stars by requiring that they all rotate at the mass-shedding limit. To do so, it is convenient to introduce the mass-shed parameter β, which provides a measure of 'how pointy' a configuration is around the equator:

$$\beta = \begin{cases} \text{spheroidal topology:} & -\left(\frac{r_e}{r_p}\right)^2 \left.\frac{d(\zeta_b^2)}{d(\varrho^2)}\right|_{\varrho=r_e} \\ \text{toroidal topology, } \varrho = \varrho_i: & \left.\frac{2\varrho_i}{\varrho_o - \varrho_i} \frac{d(\zeta_b^2)}{d(\varrho^2)}\right|_{\varrho=\varrho_i} \\ \text{toroidal topology, } \varrho = \varrho_o: & \left.\frac{2\varrho_o}{\varrho_i - \varrho_o} \frac{d(\zeta_b^2)}{d(\varrho^2)}\right|_{\varrho=\varrho_o}, \end{cases} \quad (3.51)$$

where $\zeta = \zeta_b(\varrho)$ describes the configuration's surface. This parameter is chosen such that $\beta = 0$ in the mass-shedding limit and $\beta = 1$ for (Maclaurin) spheroids.

Table 3.2. *Properties of the configuration of maximal mass for the generalized Schwarzschild class. An asterisk indicates that the corresponding quantity is a global maximum.*

Physical quantity	Value	
Gravitational mass	$M = 0.19435\ \epsilon^{-1/2}$	*
Baryonic mass	$M_0 = 0.27316\ \epsilon^{-1/2}$	*
Angular velocity	$\Omega = 1.8822\ \epsilon^{1/2}$	*
Angular momentum	$J = 0.03637\ \epsilon^{-1}$	*
Polar radius	$r_p = 0.04856\ \epsilon^{-1/2}$	
Equatorial radius	$r_e = 0.08475\ \epsilon^{-1/2}$	
Radius ratio	$A = 0.5730$	
Redshift	$z = 7.378$	*

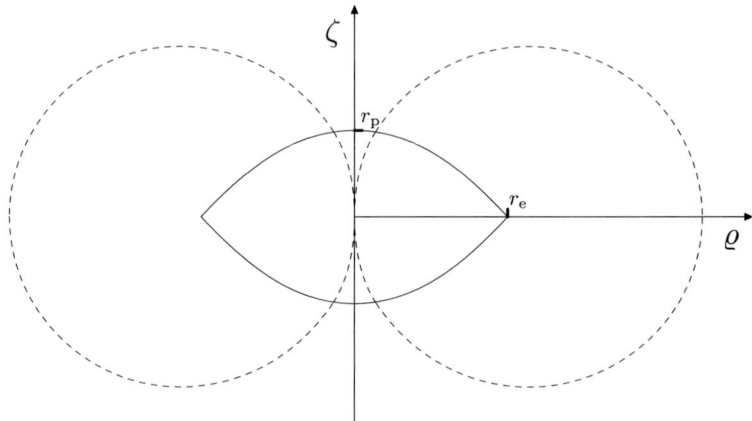

Fig. 3.16. Meridional cross-section (solid line) and ergosphere (dashed line) of the configuration of maximal mass of the generalized Schwarzschild class. Axes are scaled identically (after Schöbel and Ansorg 2003).

Let us traverse the mass-shedding sequence by choosing $\beta = 0$ and varying a second parameter, say $\Omega/\sqrt{\epsilon}$, which changes monotonically along it. We know that infinite central pressure marks one end point of the mass-shedding sequence and it turns out that the other end point is given by the Newtonian limit ((c) to (d) in Fig. 3.14). The ergosphere must shrink and vanish along this sequence since it ends in a Newtonian configuration and it turns out that the normalized angular velocity falls from $\Omega/\sqrt{\epsilon} = 1.8822$ to $\Omega/\sqrt{\epsilon} = 1.0775$ while the ratio of polar to equatorial radius falls monotonically from $A = 0.5730$ to $A = 0.1922$.

The first Newtonian lens sequence A_1^+: Although the mass-shedding sequence ends in a Newtonian limit, it clearly cannot end in a Maclaurin spheroid. There exists a Newtonian sequence connecting the mass-shedding limit to the Maclaurin solution however, which reaches it precisely at the bifurcation point A_1 ((d) to (e) in Fig. 3.14). We call this sequence the first Newtonian lens sequence because of the shape the configurations take on as they approach the mass-shedding limit. Bardeen (1971) surmised that such a sequence exists and it was studied along with other Newtonian bifurcation sequences by Ansorg *et al.* (2003c).

The Maclaurin segment: The portion of the Maclaurin sequence joining A_1 to the spherical limit comprises a Newtonian sequence that brings us back to our starting point. We have succeeded in identifying five connected physical boundary curves that thus delimit the generalized Schwarzschild class of solutions. The class itself is made of all the configurations within this boundary, the borders of which constitute its closure.

The classic Maclaurin curve, which follows from (2.15) and depicts how the square of the normalized angular velocity depends on the radius ratio A, is found in Fig. 3.17 as one of the Newtonian limits. Only three of the five boundary curves are discernable on the plot: the entire Schwarzschild sequence resides at the point $A = 1$ and the first Newtonian lens sequence A_1^+ is too small to be seen except in the inset.

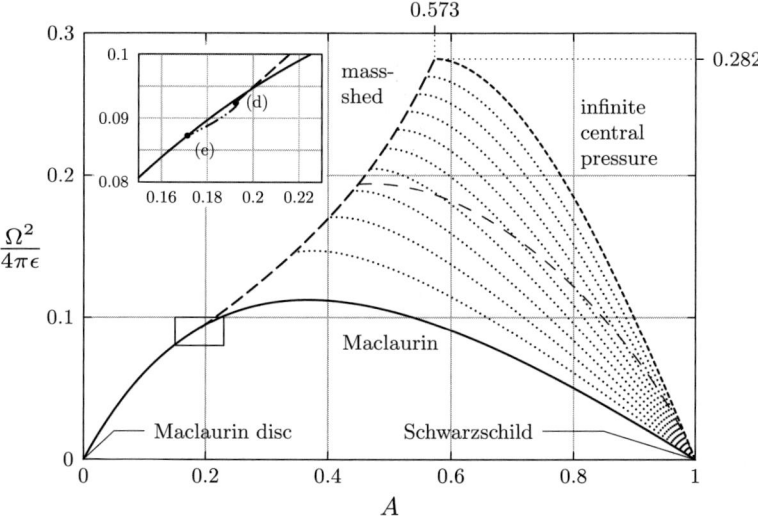

Fig. 3.17. The square of the normalized angular velocity $\Omega^2/4\pi\epsilon$ is plotted versus the radius ratio A. Ergospheres exist for configurations above the light dashed line. The labels and line types used are the same as those used in Fig. 3.14 (after Schöbel and Ansorg 2003).

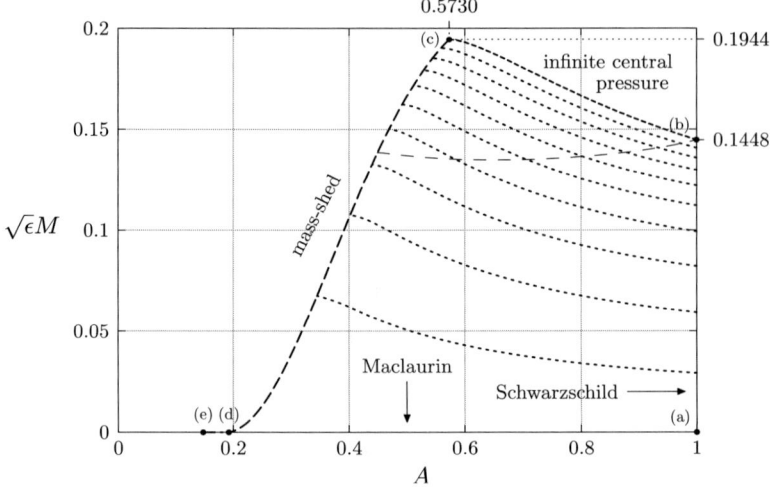

Fig. 3.18. The normalized gravitational mass is plotted versus the radius ratio. Ergospheres exist for configurations above the dashed line. The labels and line types used are the same as those used in Fig. 3.14 (after Schöbel and Ansorg 2003).

The dotted lines are lines of constant central pressure, which are equally spaced in $p_c/(\epsilon + p_c)$. The light dashed line marks the boundary between configurations with an ergosphere and those without.

In Fig. 3.18 the normalized mass is plotted versus the radius ratio. This allows us to represent the Schwarzschild sequence as a line instead of a point. In fact, all five boundary curves are visible in this plot although the A_1^+ sequence, situated between the points (d) and (e), overlaps entirely with a segment of the Maclaurin sequence. In this plot, we can see that the onset of ergospheres occurs at a roughly constant value of $\sqrt{\epsilon}M$.

3.3.3 Class II: the generalized Dyson ring class and its black hole limit

As with the generalized Schwarzschild class, the generalized Dyson ring class can be defined by locating connected boundary sequences. The members of the class and in particular the transition to a black hole were studied in Ansorg et al. (2003b). Beginning at the bifurcation point A_1, one finds that the associated Newtonian bifurcation sequence contains not only A_1^+, but also a second arm, which we denote by A_1^-. The shape of the surface along this sequence can be seen in Fig. 3.19. The first three members in this sequence of pictures belong to A_1^+ and were discussed in the previous section. The subsequent ones belong to the bifurcation sequence

3.3 Equilibrium configurations of homogeneous fluids 145

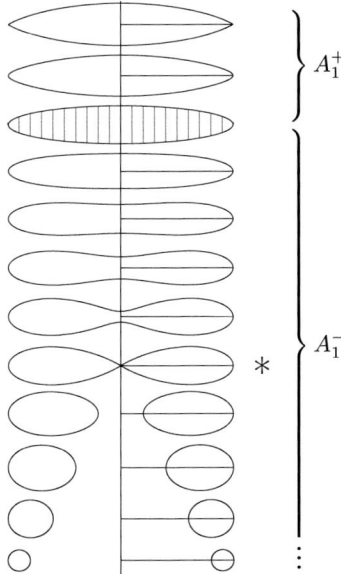

Fig. 3.19. Meridional cross-section of Newtonian bodies belonging to the first bifurcation sequence with ζ/r_e (ζ/ϱ_o) plotted versus ϱ/r_e (ϱ/ϱ_o). The portion of the sequence running from the Maclaurin spheroid (denoted by hatching) to the mass-shedding limit belongs to A_1^+ and the other portion to A_1^-. The transition point from spheroidal to toroidal topology is marked by an asterisk (after Ansorg et al. 2003b).

A_1^- ((e) to (f) in Fig. 3.14), which is a Newtonian boundary curve of the class of solutions being discussed here. The original ellipsoid grows increasingly dumbbell shaped and finally pinches together at the centre and takes on a toroidal topology.

For toroidal objects, a numerical code using coordinates like those described in Subsection 3.2.2 is needed. The configuration at the transition point provides possible 'initial data' for the program as does the Newtonian thin ring limit to be discussed below. As always, once one has succeeded in finding a single numerical solution, it can be used as the initial guess in the Newton–Raphson method for determining the solution to a neighbouring configuration.

Newtonian rings received considerable attention in the nineteenth century and a good deal is known about them. In particular, Dyson (1892, 1893) determined the gravitational potential by carrying out an expansion to fourth order about the thin ring limit in which the ring's cross-section becomes circular. The results are remarkably accurate even when A deviates significantly from -1 (see e.g. Fig. 1 in Ansorg et al. 2003c). We see in Fig. 3.19 that the rings in this sequence are tending toward the thin ring limit. In fact, A can be arbitrarily close to -1. In that limit,

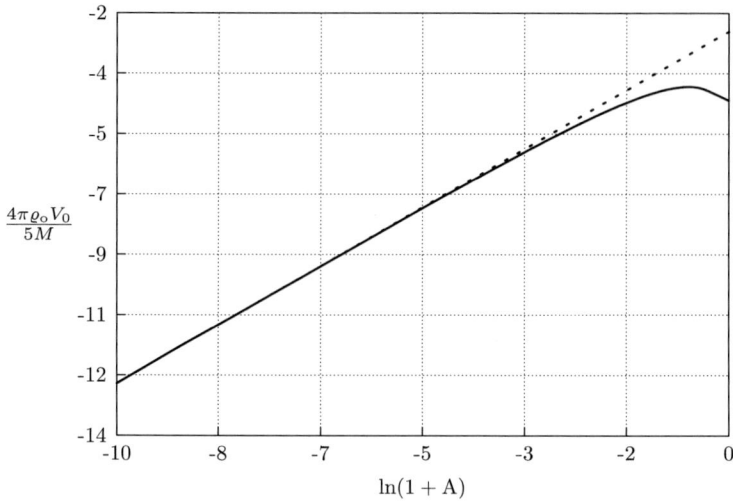

Fig. 3.20. A plot demonstrating the behaviour of Newtonian rings in the thin ring limit $A \to -1$. As the function plotted here tends to $-\infty$, it approaches the straight line given by Equation (3.52) and denoted by a dotted line.

if the extent of the ring ϱ_0 and its mass M are kept finite, it turns out that V_0 as a function of A tends logarithmically to minus infinity such that

$$\lim_{A \to -1} \frac{4\pi \varrho_0 V_0}{5M} - \ln(1+A) = \frac{1}{4} - \ln 16 \qquad (3.52)$$

holds (Dyson 1892, Horatschek and Petroff 2008; see also Equation (11) in Fischer *et al.* 2005). A graphical depiction of this behaviour can be found in Fig. 3.20. Since Newtonian theory is obtained as $V_0 \to 0$, we see that this thin ring limit poses a consistency problem. We shall return to this interesting point after considering the other boundary sequences.

Moving back to the point A_1, we can traverse the Maclaurin sequence to the next bifurcation point A_2 ((e) to (i) in Fig. 3.14). Bifurcating from this point is the A_2^+ sequence, which is very similar to A_1^+. It too ends in a mass-shedding limit after a short time, short meaning that A, Ω and J vary only slightly along it ((i) to (h) in Fig. 3.14). Travelling now along the mass-shed sequence, we find that the surface of the body pinches in at the centre as we depart from the Newtonian limit. There is a change in topology, which occurs at $1 - e^{V_0} \approx 0.2$. As we can see in Fig. 3.21, the value of this parameter can approach the value of one, i.e. the relative redshift can tend to infinity for rings with a radius ratio less than $A \approx -0.56$ ((h) to (g) in Fig. 3.14). As indicated in the figure, this is the limit in which the exterior metric becomes that of the extreme Kerr black hole (see Subsection 1.8.2 for a discussion of the exterior metric and the 'inner world').

3.3 Equilibrium configurations of homogeneous fluids

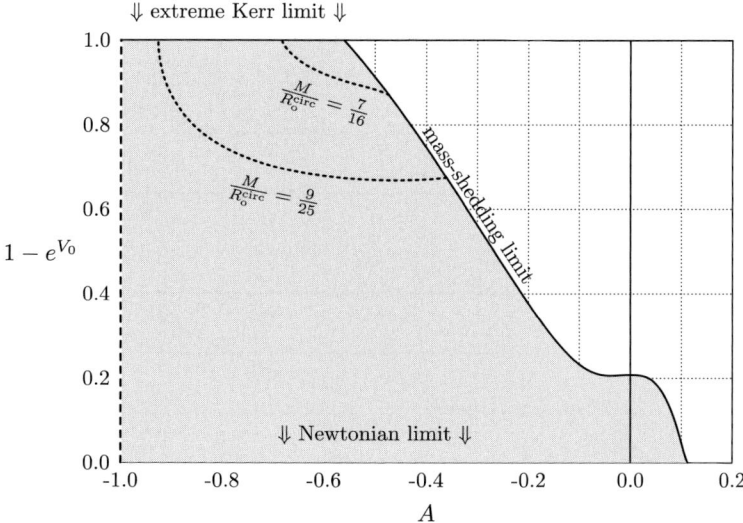

Fig. 3.21. In this plot of $1 - e^{V_0}$ versus A, the mass-shedding sequence marks a boundary for the configurations of Class II, which lie entirely within the shaded region. The line indicating the transition from a spheroidal to a toroidal topology is emphasized.

Let us consider the evidence supporting this claim. We recall that Meinel (2006) has shown that a necessary and sufficient condition for the existence of a parametric transition to an extreme Kerr black hole is that $e^{V_0} \to 0$. It is not to be expected that it will be easy to come arbitrarily close to this limit using numerical means. After all, it becomes increasingly necessary to resolve two increasingly disjoint worlds. For a ring quite close to this limit, the metric functions in its vicinity resemble those of a non-extreme ring. The asymptotic behaviour of the 'inner world' describes the functions' behaviour far away from the ring, but these functions must possess large gradients in order to assume their asymptotically flat values. Making use of logarithmic coordinates such as those described by (3.15), it was nonetheless possible to find rings with $e^{V_0} \approx 10^{-3}$. In light of the proofs mentioned above, this alone is convincing support of our claim. We choose to go further however and explore the behaviour of the metric. Examining multipole moments, it was possible to show that they indeed approach those of the extreme Kerr metric (Labranche *et al.* 2007). Moreover, they tend to these values in a way that is independent of the equation of state and independent of whether or not the transition body is a ring or a disc of dust. There also exists evidence showing that the metric in the vicinity of the ring approaches that of the inner world. The sequence of plots in Fig. 3.22 shows the development of the throat geometry and is inspired by Fig. 13 in Bardeen and Wagoner (1971). The final plot in the sequence shows two separate worlds after

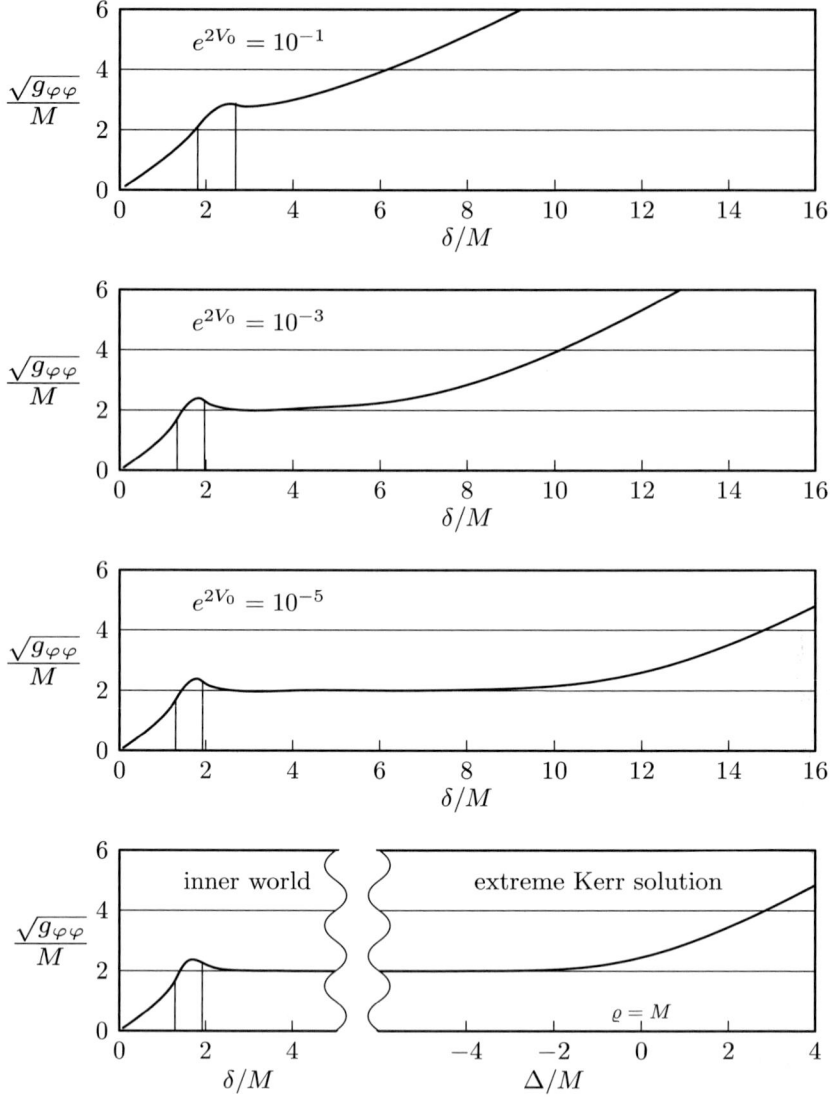

Fig. 3.22. The function $\sqrt{g_{\varphi\varphi}}$ in the equatorial plane is plotted versus proper radial distance $\delta = \int_0^\varrho \sqrt{g_{\varrho\varrho}}\,d\varrho$, both normalized with respect to the mass M, for a sequence approaching the extreme Kerr limit $e^{2V_0} = 0$. All four plots were made for a homogeneous ring with a radius ratio $A = -0.7$ and with a value for e^{2V_0} as indicated. The thin vertical lines indicate the positions of the inner and outer radii of the ring. In the last plot ($e^{2V_0} = 0$), Δ gives the proper radial distance in the Kerr metric to the reference point $\varrho = M$. In the throat region, $\sqrt{g_{\varphi\varphi}}/M$ tends to the constant value 2, cf. page 30. Note that the proper distance between any point in the 'inner world' region and any point in the 'extreme Kerr' region tends to infinity as $e^{2V_0} \to 0$, cf. (1.147) (after Labranche et al. 2007).

3.3 Equilibrium configurations of homogeneous fluids

having carried out the limiting process. To construct the inner world solution, a spectral program was used, which prescribes the correct non-flat asymptotic values for the functions as discussed in Subsection 1.8.2. The startup data that we used for this program were provided by simply cutting off the far exterior region for rings from the asymptotically flat program for rings with $e^{V_0} \ll 1$. The fact that such startup data work, provides additional evidence suggesting that the metric in the ring's vicinity approaches that of the inner world.

Let us return now to the point (f) in Fig. 3.14, marked as an open circle, where the extreme Kerr and the Newtonian A_1^- sequence approach each other. From the standpoint of Fig. 3.21, this amounts to representing the whole line $A = -1$ by this open circle. This symbolic representation has its origin in the singular nature of the thin ring limit $A \to -1$.

The physical requirements that one can impose on a ring for finite V_0 are that its mass and its extent remain finite. As an invariant measure of the extent of the ring, we here choose to use the circumferential radius of the outer edge in the equatorial plane R_o^{circ}, which is simply the proper circumference at $\varrho = \varrho_o$, $\zeta = 0$ divided by 2π. Based on the Newtonian result (3.52), one might suppose that for a finite fixed value of V_0, $M/R_o^{\text{circ}} \to 0$ as $A \to -1$. Figure 3.23 provides an example of the verification of this expectation for $e^{V_0} = 0.6$. The constant $4\pi V_0/5$ is not of particular importance in this example and was simply chosen since the curve is then known to be linear in the Newtonian limit. What is significant is that indeed $M/R_o^{\text{circ}} \to 0$ in the thin ring limit making it a physically unacceptable model. This explains the reason for choosing a dashed line for $A = -1$ in Fig. 3.21. Two sequences with constant, finite M/R_o^{circ} can be found in that figure. Each

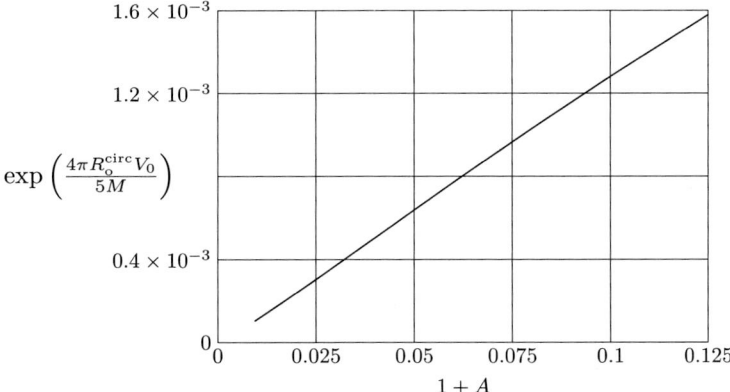

Fig. 3.23. The function $\exp\left(\frac{4\pi R_o^{\text{circ}} V_0}{5M}\right)$ is plotted versus $1 + A$ for a sequence with $\exp(V_0) = 0.6$, cf. Fig. 3.20 and Equation (3.52).

such sequence runs from the mass-shedding limit to that of the extreme Kerr black hole. The most relevant such curve in the current context would have a value $M/R_0^{\text{circ}} = \text{constant} \ll 1$. For configurations sufficiently far away from the thin ring limit, this condition indicates the validity of the Newtonian limit. Thus the corresponding curve in Fig. 3.21 would 'begin' along the mass-shedding curve arbitrarily near the Newtonian limit $1 - e^{V_0} = 0$. As such a curve nears $A = -1$, Equation (3.52) tells us that $|V_0|$ would grow rapidly, which turns out to hold even far away from the Newtonian limit. The curve would thus make a sharp bend upward and follow the line $A = -1$ right up to its end point at $1 - e^{V_0} = 1$.

3.3.4 Class III: the first generalized core-ring class

The configurations comprising the Newtonian bifurcation sequence associated with A_2 and belonging to the third class begin to take on a rolling-pin shape in cross-section as they depart from the spheroid. This effect becomes more pronounced as one proceeds along sequence A_2^- ((i) to (j) in Fig. 3.14) until a ring appears which is on the verge of separating from a central body. We consider such configurations to mark a boundary in the configuration space, since our intention here is to classify *single* homogeneous bodies. Thus configurations at the transition point separating single bodies from two bodies form a boundary sequence running from the Newtonian limit to the mass-shedding limit ((j) to (k) in 3.14). As one moves away from the Newtonian limit, $\Omega/\sqrt{\epsilon}$ increases along this sequence until shortly before reaching the mass-shedding limit, when it again falls slightly. Its global maximum in this class is found at the bifurcation point A_2 however. Along the mass-shed sequence ((k) to (l) in Fig. 3.14), the configurations again become rolling-pin shaped before finally resembling a spheroid with a cusp at the equator in the Newtonian limit. The Newtonian A_3^+ sequence, which can be seen in Fig. 3.24, then takes us to the true spheroid at the bifurcation point A_3 ((l) to (m) in Fig. 3.14), which is then connected to A_2 via the Maclaurin sequence ((m) to (i) in Fig. 3.14). Having again come full circle, we have identified a closed circuit of boundary sequences defining the borders of this class of solutions.

3.3.5 Class IV: the first generalized two-ring class

As one traverses the Newtonian A_3^- sequence ((m) to (n) in Fig. 3.14), the highly flattened stars develop two bulges. Figure 3.24 depicts the surface of the configurations in the third bifurcation sequence. The A_3^+ portion of it, which belongs to class III, runs from the hatched Maclaurin spheroid to the mass-shedding star at the top of the figure. The sequence we are concerned with runs downward from the Maclaurin spheroid. Along the way there is a change in topology and

3.3 Equilibrium configurations of homogeneous fluids

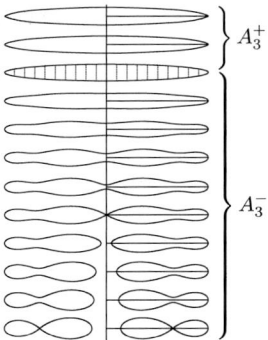

Fig. 3.24. Meridional cross-section of Newtonian bodies belonging to the third bifurcation sequence with ζ/r_e (ζ/ϱ_0) plotted versus ϱ/r_e (ϱ/ϱ_0). The portion of the sequence running from the Maclaurin spheroid (denoted by hatching) to the mass-shedding limit belongs to A_3^+ and the other portion to A_3^- (after Ansorg *et al.* 2003c).

one finds rings with a bowling pin shape in cross-section. The tapering in this 'bowling pin' becomes more pronounced until one finally reaches a ring on the verge of the transition to two rings. Such two-body configurations again mark a boundary of our solution space and can be followed from the Newtonian limit to the mass-shedding limit ((n) to (o) in Fig. 3.14). Along the mass-shedding sequence ((o) to (p) in Fig. 3.14) the figures again pass through the transition point from toroidal to spheroidal topologies and one ends in a Newtonian limit that is barely distinguishable from a spheroid. For example, along the entire Newtonian A_4^+ sequence ((p) to (q) in Fig. 3.14), the relative change in physical parameters such as A, Ω or J is only a few percent. This boundary sequence brings us back to the Maclaurin solutions at the point A_4. The final boundary sequence enclosing the solutions in class IV is made up of the spheroids in the Newtonian segment joining bifurcations points A_3 and A_4 ((q) to (m) in Fig. 3.14). Relative changes in the parameters listed above are about a factor of five greater along this segment than along the A_4^+ sequence.

3.3.6 Overview of the solution space including the disc limit

As was explained in Subsection 3.3.1, it is possible to divide the solution space of rigidly rotating, relativistic, homogeneous bodies into a countably infinite number of classes of solutions (for a detailed discussion, see Ansorg *et al.* 2004). The first three of them are depicted in Fig. 3.25, which constitutes a fuller version of Fig. 3.17. Each class contains a segment of the Maclaurin solution between adjacent bifurcation points as one of its boundaries and is comprised of all configurations

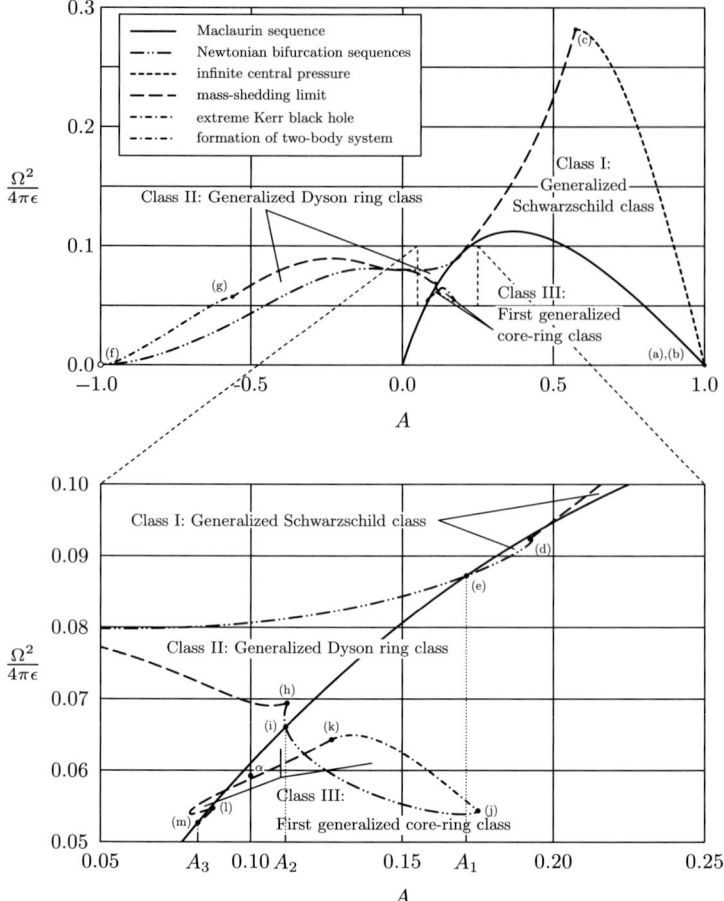

Fig. 3.25. The square of the normalized angular velocity $\Omega^2/4\pi\epsilon$ is plotted versus the radius ratio A for the first three classes. A blowup containing Class III provides greater detail (after Ansorg *et al.* 2004).

that can reach this boundary via a continuous transition. It turns out that each class has five boundary sequences enclosing it and we refer again here to the schematic overview in Fig. 3.14, which has played an important role throughout Section 3.3 in discussing individual classes.

The first two classes are exceptional in that each contains boundary sequences unique to that class. The generalized Schwarzschild class (Class I) contains the static limit and that of infinite central pressure. Every other class contains a finite limit for the maximal pressure. The generalized Dyson ring class (Class II) allows for a transition to the extreme Kerr black hole, something otherwise possible only after reaching the disc limit. To reach this limit, an infinite number of further classes must be traversed. Each of them contains three Newtonian boundary sequences: a

Maclaurin segment, a bifurcation sequence ending in a mass-shedding limit and a bifurcation sequence ending in a two-body configuration. In each of these higher classes, a two-body and simultaneously mass-shed configuration exists, meaning that the class is indeed bounded by five sequences. This configuration is likely to be the one with the highest relative redshift in the class. The higher classes can be further divided into even and odd ones. The two-body limit in the $(2n+1)$-st class consists of a central core with n humps surrounded by a ring just barely touching it. Similarly, the two-body limit in the $(2n+2)$-nd class consists of a ring with n humps surrounded by a ring just barely touching it. These statements are based on following the behaviour of the first ten Newtonian bifurcation sequences (see Ansorg *et al.* 2003c), a detailed study of the first four relativistic classes and then extrapolating.

As one proceeds to ever higher classes, the configurations grow increasingly flat and closely resemble a many-ring system. Entire classes deviate only marginally from the Newtonian limit as measured by the maximum value for the relative redshift for example. As $n \to \infty$, one approaches the Maclaurin disc limit. This boundary configuration is simultaneously the Newtonian limit of the rigidly rotating disc of dust.

The solution space presented here includes all homogeneous figures of equilibrium with a Newtonian limit. It may well be the entire solution space. It is difficult to rule out the possibility that there exist relativistic solutions that do not possess a Newtonian limit whatsoever. Such solutions, if they exist, could be remote islands in some parameter space that are difficult to chance upon with the numerical methods used here.

3.4 Configurations with other equations of state

Figures of equilibrium of constant density have played a crucial historical role and are of particular academic interest because of their tractability to analytic methods. Models with other equations of state are of greater astrophysical relevance however, and we thus want to touch on some of their properties here.

Arbitrary barotropic equations of state present no significant difficulties to the numerical methods presented in this book. Indeed the inaccuracies inherent to mass-shedding configurations are less pronounced for equations of state without discontinuities in the density ϵ at the surface of the body, as explained in Section A1.1. Equations of state involving phase transitions may require the introduction of additional computational domains and tabulated ones the introduction of (ideally analytic) functions $p = p(\epsilon)$, but these are minor technical points that can be addressed as the need arises. Here we shall compare homogeneous bodies to those governed by three other equations of state: various polytropic ones with particular emphasis on the one with the polytropic index $n = 1$, the one describing a completely degenerate ideal Fermi gas of neutrons and one describing strange matter based on the MIT bag model. Each of these equations of state was discussed in Section 1.5. The governing principle that will be used in this subsection for deciding which exemplary configurations to bring, is to restrict ourselves to the first two classes of solutions and to provide a rough overview of the properties of the first of them.

3.4.1 Polytropic equations of state

The polytropic equation of state

$$p = K\mu_B^\gamma = K\mu_B^{1+\frac{1}{n}} \tag{3.53}$$

provides a mathematically simple relation between the pressure p and the baryonic mass density μ_B, where K is known as the polytropic constant. The polytropic index n determines the 'stiffness' of the equation, which can range from 0 for the stiffest case, homogeneous bodies, to large values for 'soft' equations of state, which exhibit a dense concentration of matter surrounded by a tenuous envelope. In Newtonian theory, in which μ_B simply becomes the mass density, closed form solutions are known in the static case for $n=0$, $n=1$ and $n=5$, the last value also being the maximal attainable one for spherically symmetric Newtonian polytropes.

When using the methods presented in this book for the numerical description of a polytropic configuration, one has to keep in mind that $\epsilon(h)$ is an analytic function of h at $h(0)$ only for $n = 0, 1, 2, \ldots$, a property which also carries over to the metric functions. As was mentioned in the footnote on page 128, this affects the rate of convergence. Nonetheless, generic configurations with non-integer n can be calculated to high accuracy without great numerical effort. An example of the convergence rate for a chosen sequence as it depends on n is provided in Fig. 3.26. Using the measure R_ϕ^m as defined in (3.34), one sees that the solutions for small values of the spectral resolution m are extremely accurate when $n=0$ or $n=1$. For other values of n, one can choose m to be well in excess of 40 if one requires extremely high accuracy, though $m=15$ is sufficient for most purposes.

Having come to know the nature of the solution space for homogeneous bodies in the last chapter, it is natural to wonder how a change in the equation of state affects that picture. We begin addressing that question by looking at what happens in the vicinity of the bifurcation point A_1 when the polytropic index n is perturbed slightly. In the homogeneous case, it turned out that the Newtonian bifurcation

3.4 Configurations with other equations of state

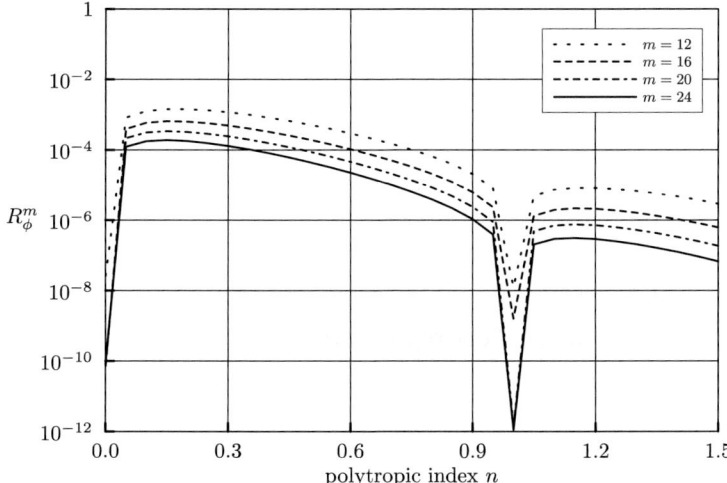

Fig. 3.26. Convergence is shown in dependence of the polytropic index n for a sequence with constant radius ratio $A = 0.8$ and constant central pressure $K^n p_c = 10^{-4}$ ($p_c/\epsilon = 1$ for $n = 0$). The index m indicates how many Chebyshev polynomials were used in each dimension of each domain and the measure of convergence, R_ϕ^m, is defined in (3.34). The polytropic index n was varied in steps of 0.05, which thereby determines the width of the spikes at $n = 0$ and $n = 1$. In the continuum limit, these spikes would be much narrower.

points are singularities in the post-Newtonian expansion and that configurations with radius ratio $A = A_k$ and mass-shed parameter $\beta = 1$ can only be reached in the Newtonian limit. Similarly, it seems that polytropic Newtonian configurations can only reach such points for $n \to 0$. The behaviour of three Newtonian sequences in the vicinity of A_1 is depicted in Fig. 3.27 for $n = 10^{-2}$, $n = 10^{-3}$ and $n = 10^{-4}$. Comparison with Fig. 3.15 shows that the progression for homogeneous bodies from the Newtonian limit to slightly relativistic ones is very similar to the progression in the Newtonian limit from polytropes with $n = 0$ to slightly larger values.

The similarities between Figs 3.15 and 3.27 suggest that for polytropes (with $n > 0$) it will also be possible to identify disjoint classes of solutions. Here, two 'adjacent classes' are not going to have a point in common however. We no longer have an anchoring sequence such as the Maclaurin sequence, binding together the classes and providing a natural means of ordering them linearly. It will still be the homogeneous classes that we use for providing an ordering scheme and a nomenclature for a more general solution space, however. To that end let us start by calling a configuration a basis configuration if it is related to the homogeneous bodies located at the points (b), (f), (j), (n), etc. of Fig. 3.14 by remaining at the intersection of the two pertinent sequences (e.g. two-body and Newtonian for the point (j)) and by continuous variation of the polytropic index n. For a

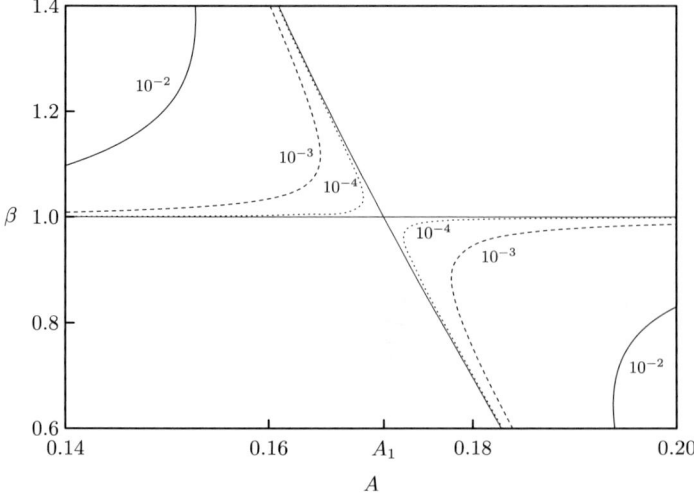

Fig. 3.27. Newtonian sequences in the vicinity of the bifurcation point A_1 are depicted for polytropes with the indicated value of the polytropic index n. See text for a comparison with Fig. 3.15.

given polytropic equation of state, the k-th class can then be defined to consist of those configurations connected via continuous parameter variations to the basis configuration associated with that class.

Having introduced this broader notion of a class of solutions, let us proceed to look at the first two classes for polytropes with $n = 1$. The choice of this value for the polytropic index is somewhat arbitrary, but has the advantages of being fairly widely discussed in the literature for one, and being sufficiently far away from $n = 0$ so as to have some qualitatively new features for another.

Polytropes with $n = 1$

The first class of solutions for polytropes with $n = 1$ is enclosed by four boundary sequences: static, infinite central pressure, mass-shed and Newtonian. One sequence is missing compared to the five homogeneous ones since the Newtonian limit is no longer divided into segments by a bifurcation point. The static limit is the most accessible and best understood limit, merely requiring that one solves the TOV equation (see Section 1.7.2). As in the homogeneous case, the sequence of static polytropes with $n = 1$ also runs from the Newtonian limit to that of infinite central pressure. The Newtonian solution, as was mentioned above, is analytically known. Starting from this limit and choosing central pressure to parameterize the sequence, one finds for a given equation of state (i.e. for a constant value of K) that gravitational mass is no longer a monotonically increasing function. The behaviour of mass as it depends on the circumferential radius is depicted in Fig. 3.28. There is

3.4 Configurations with other equations of state

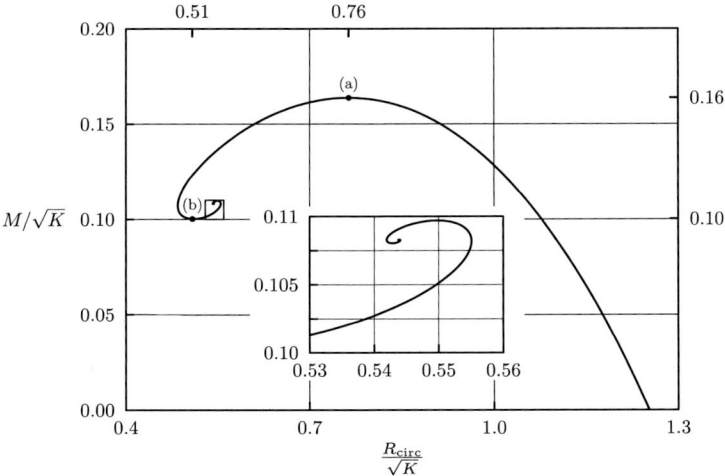

Fig. 3.28. The mass of static polytropes with $n = 1$ is plotted as a function of circumferential radius $R_{\text{circ}} = \sqrt{g_{\varphi\varphi}(\varrho = r_e, \zeta = 0)}$. The points labelled (a) and (b) are discussed in the text.

a pronounced global maximum and there are many local maxima as one spirals in toward the limit of infinite central pressure. The global maximum, which has been labelled '(a)', also marks a maximum for the baryonic mass and the onset of an instability with respect to axisymmetric perturbations as discussed in Chapter 4. A further point of interest, '(b)', marks the transition point from positive to negative binding energies $M_0 - M$. It is well into the unstable branch of solutions and is further evidence indicating that this branch is not astrophysically relevant. Negative binding energies in general relativity have been discussed in the context of static polytropes by Tooper (1964).[4] Although in the example provided here, the binding energy becomes negative far away from the Newtonian limit, the phenomenon as such is not solely relativistic. The binding energy of Newtonian static polytropes with $n > 3$ is known to be negative for example.

The infinite central pressure sequence runs from the static to the mass-shedding limit. This sequence is not as relevant as the corresponding one for homogeneous configurations since physical parameters such as mass, redshift and angular velocity are maximal for finite values of the central pressure. Furthermore, the binding energy remains negative over the whole of the sequence.

Joining on to the sequence of infinite central pressure is the mass-shedding sequence. It is more astrophysically relevant, not least of all because the maximal values for mass, baryonic mass, redshift, angular momentum and angular velocity

---

[4] Note that 'polytrope' in Tooper (1964) is taken to be $p = K\epsilon^\gamma$ and not $p = K\mu_B^\gamma$ [see (1.46)].

Table 3.3. *Maximal values for physical quantities are provided for polytropes with n = 1, where 'maximal' refers here to a global maximum for Class I.*

Physical quantity	Value	A
Gravitational mass	$M_{\text{static}} = 0.164 \quad K^{1/2}$	1.000
Gravitational mass	$M = 0.188 \quad K^{1/2}$	0.585
Baryonic mass	$M_0 = 0.207 \quad K^{1/2}$	0.585
Angular velocity	$\Omega = 0.634 \quad K^{-1/2}$	0.603
Angular momentum	$J = 0.0204 \quad K$	0.583
Redshift	$z = 0.529$	0.596

are all to be found along it. A list of such values can be found in Table 3.3. The most striking difference between homogeneous stars and polytropes with $n = 1$ is that the maximal redshift of the latter is a factor of 14 smaller. Because redshift depends so strongly on the stiffness of the equation of state, it could provide important clues as to the nature of matter under extreme conditions.

The Newtonian limit, which marks the second end point of the mass-shed sequence, brings us full circle back to the static case. Class I for polytropes with $n = 1$ has thus been staked out and can be seen along with the second class in Figs 3.29 and 3.30. In the first of these figures, one can see that the mass-shed sequence, starting from the Newtonian limit, reaches a minimum value along the y-axis, curves upward, 'overshooting' the $p_c = \infty$ mark, before doubling back on itself again (a process which, like the inward spiral in Fig. 3.28 may be repeated *ad infinitum*). Although the boundary curves are depicted, the possible solutions in Class I do not all reside within these curves in the parameter space chosen in Fig. 3.29. Consider, for example, the sequence generated by holding the value for the central pressure constant such that it meets the mass-shedding sequence at its smallest value for $M^2\Omega^2/z^3$. The curve described by this sequence lies below that of infinite central pressure. In contrast, Class I configurations do all lie within the boundary curves of Fig. 3.30. The impression of a shrinking area for Class I that one gains by comparing this figure with Fig. 3.25 is thus not entirely unwarranted: All Class I stars for polytropes with $n = 1$ have a radius ratio greater than 0.5 whereas this ratio can be less than 0.2 for homogeneous stars.

Turning our attention now to the second class, one can see that the gap between it and the first class, which was illustrated in Fig. 3.15, is very pronounced in Figs 3.29 and 3.30. In fact, the entirety of the second class is made up of ring configurations, there being no transition point from toroidal to spheroidal topologies for polytropes with $n \gtrsim 0.36$ (see also Fig. 3.35 below). As with the homogeneous second class,

3.4 Configurations with other equations of state

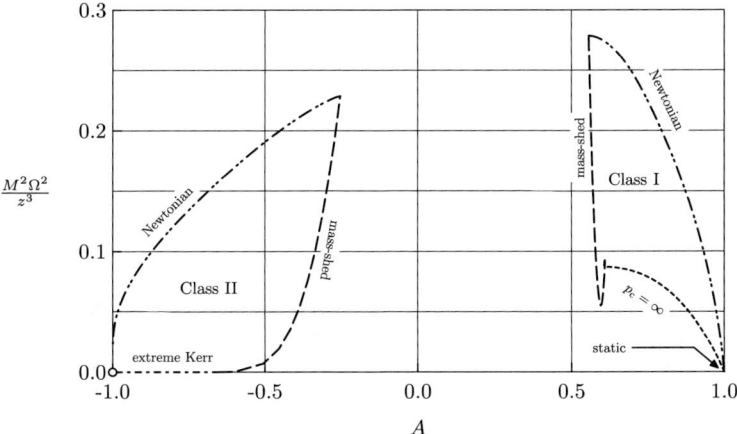

Fig. 3.29. The boundary curves for the first two solution classes for polytropic configurations with $n = 1$. The dimensionless parameter $M^2\Omega^2/z^3$ was chosen because it allows a representation of both the Newtonian sequence and that of infinite central pressure. The signification of the line types is given in the legend of Fig. 3.25.

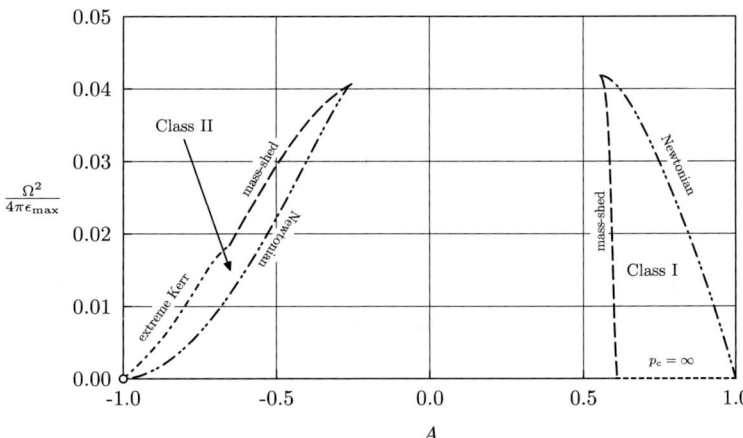

Fig. 3.30. The boundary curves for the first two solution classes for polytropic configurations with $n = 1$. This plot can be compared with Fig. 3.25, the legend of which also clarifies the meaning of the line types.

the limiting curves are here too made up of Newtonian, mass-shed and extreme Kerr sequences. The single (non-bifurcating) Newtonian curve means that there are a total of three as opposed to five boundary sequences however. Newtonian polytropic rings with $n = 1$ can be described analytically by making use of an expansion about the thin ring limit (Ostriker 1964, Petroff and Horatschek 2008), but become increasingly inaccurate as one approaches the mass-shedding end of the sequence. The mass-shedding sequence runs between the Newtonian and the extreme Kerr limit. This third limiting sequence, for which the 'interior' and non-asymptotically flat ring metric separates from the 'exterior' extreme Kerr metric, completes the triangle.

The solution space for polytropes with arbitrary indices

Acquiring as detailed a picture of the solution space for polytropic bodies as was possible for homogeneous ones would require an enormous amount of work. Here we merely intend to list some of what is known in order to gain an understanding of some of it. For one thing, we know from static Newtonian theory that polytropic stars with $n > 5$ do not exist and can assume that relativistic configurations, at least in a neighbourhood of this limit, do not exist either. We know further that for $n > 3$ the binding energy is negative, thus shedding light on their astrophysical relevance. It is interesting to note that rings with $n > 3$ do have positive binding energies however.

As was mentioned on page 155, the transition from $n = 0$ to higher n in the Newtonian limit is similar to an increase in some relativistic parameter. In particular, the pinching together that is seen along Newtonian bifurcation sequences can also be found by increasing the polytropic index. An example of the shape of such a configuration can be found in Fig. 3.31. What is apparent in this picture, is that the pinching together takes place at the innermost lobe as opposed to the outermost one as is the case for the homogeneous bifurcation sequences.

Starting from a 'spheroidal-like' basis configuration and increasing n leads to a configuration that is both at the mass-shedding limit and at the two-body transition limit. For the third class, this double limit is reached for $n \approx 1.2$ meaning that there exists no third class for $n \gtrsim 1.2$. As one proceeds to higher odd classes, the value of n for which one reaches the double limit decreases slightly and has fallen to a value of $n \approx 1.1$ by the time one has reached Class XXI. It seems likely that similar behaviour is exhibited for ring topologies, so that one can conjecture that a finite number of classes exist for polytropes with $n > n_0$, where the index n_0 may or may not be arbitrarily small, but where the upper limit $n_0 < 1.1$ holds.

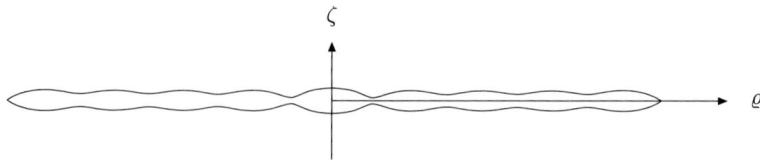

Fig. 3.31. Meridional cross-section of the Newtonian mass-shedding configuration from Class IX with $n = 1$.

3.4.2 Completely degenerate, ideal gas of neutrons

In order to present numerical results for an ideal neutron gas, it will be useful to make use of dimensionless quantities and we shall thus introduce an appropriate constant. Naming the constant of Equations (1.48)

$$\tilde{K}_n := \frac{m_n^4}{24\pi^2 \hbar^3}, \tag{3.54}$$

the series expansion for the pressure of (1.48a) about $x = 0$ reads

$$p = \tilde{K}_n \left(\frac{8}{5} x^5 + \mathcal{O}(x^7) \right). \tag{3.55}$$

In the limit $x \to 0$, the relation

$$p = \frac{\tilde{K}_n^{-2/3}}{20} \mu_B^{5/3} \tag{3.56}$$

then follows. Introducing

$$K_n := \frac{1}{20} \tilde{K}_n^{-2/3}, \tag{3.57}$$

a constant has been defined that will be used to render quantities dimensionless such that, in the Newtonian limit, the values for physical quantities for the completely degenerate neutron gas tend to the dimensionless analogues for a polytrope with $n = 3/2$. The constant K_n has, of course, the same unit as the polytropic constant K associated with polytropes with $n = 3/2$.

We now present a brief overview of the first class of solutions and shall mention the second one in the context of Subsection 3.4.3. The precise definition of a class of solutions for a completely degenerate, ideal Fermi gas follows naturally from the definition for polytropes. Because the basis configurations introduced on page 155 are located at the Newtonian limit, the equation of state being considered here coincides with that of the polytropes. The set of configurations connected via a continuous transition to a given basis configuration is then said to comprise the associated class of solutions.

The qualitative features of the first class of solutions are much like those of polytropes with $n = 1$. The four boundary sequences are once again the static one, that of infinite pressure, the mass-shedding and the Newtonian sequence. A plot of the square of the angular velocity divided by the central density as it depends on radius ratio can be found in Fig. 3.32 for the boundary curves. Here some of the detail exhibited by the mass-shedding sequence as one approaches the curve of infinite central pressure is provided by the blowup. As a point of technical interest, it should

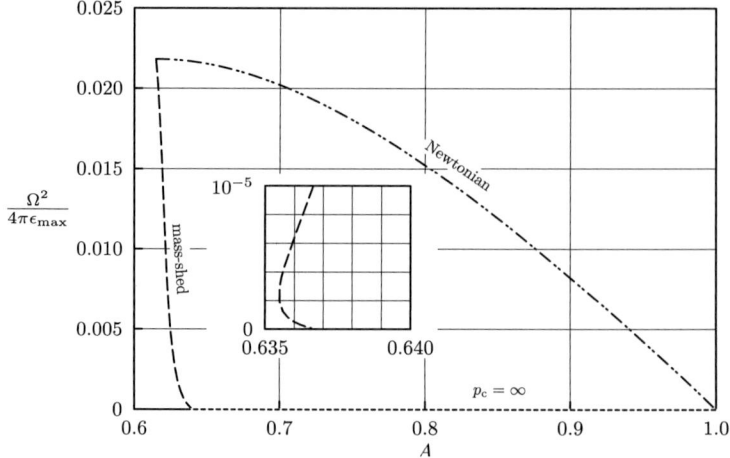

Fig. 3.32. The boundary curves for the first solution class for configurations made up of a completely degenerate, ideal gas of neutrons. The blowup in the centre shows some of the detail of the curves in the vicinity of infinite central pressure. This plot can be compared with Fig. 3.30.

be noted that it is necessary to span fourteen orders of magnitude in central pressure in order to produce Fig. 3.32. As with polytropes, the values for mass, angular velocity and redshift do not increase monotonically as one increases central pressure along sequences with constant angular momentum. They are however again maximal along the mass-shedding curve and some values are listed in Table 3.4. Unlike the constants of other equations of state discussed in this book, K_n of (3.57) is fixed and thus enables us to provide dimensional values for physical quantities without having to make somewhat arbitrary choices.[5] One finds, for example, that this model is capable of accounting for observed rotational periods, but not observed masses for neutron stars, which exceed $0.78\,M_\odot$. The maximal redshift is quite small compared to that of homogeneous stars and the increase in maximal mass due to rotation is only about 10%. A much greater value for this increase in maximal mass is found for the MIT bag model to be discussed in the next subsection.

3.4.3 Strange Matter

The strange matter equation of state,

$$\epsilon = 3p + 4B, \tag{3.58}$$

---

[5] It may be argued that the value for the MIT bag constant B of Equation (1.50) follows from the model within fairly narrow margins. Since other models for strongly interacting matter can lead to equations of state with the same form, but radically different values for the constant, we choose not to specify its value further (see page 163 for a reference to other derivations of equations like (1.50)).

3.4 Configurations with other equations of state

Table 3.4. *Extremal values for physical quantities are provided for configurations made up of a completely degenerate, ideal gas of neutrons, where 'extremal' refers here to a global maximum (minimum for the rotational period) for Class I.*

Physical quantity	Value		A
Gravitational mass	$M_{\text{static}} = 0.236\, K_n^{-3/4}$	$= 0.710\ M_\odot$	1.000
Gravitational mass	$M = 0.260\, K_n^{-3/4}$	$= 0.782\ M_\odot$	0.620
Baryonic mass	$M_0 = 0.270\, K_n^{-3/4}$	$= 0.812\ M_\odot$	0.620
Rotational period	$2\pi/\Omega = 0.196\, K_n^{-3/4}$	$= 0.476$ ms	0.629
Redshift	$z = 0.216$		0.624

based on the MIT bag model and introduced in (1.50), is a conceivable model for quark matter in a deconfined state and presents an alternative to more traditional models for matter in dense stars. The same equation of state was considered for entirely different reasons by Sen (1934), who applied it to static stellar models. A detailed study of much of what we call the first class was made by Gourgoulhon *et al.* (1999), who also provide references to earlier work on the subject. Similar equations of state, but with different values for the proportionality factor between ϵ and p, and with different expected values for the constant B, have been derived by different means, see e.g. Peshier *et al.* (2000).

One can see that at a configuration's surface $p = 0$, the energy-density has a discontinuity, as in the homogeneous case. Furthermore, in the Newtonian limit $p/\epsilon \to 0$ the equation of state reduces to $\epsilon = $ constant, meaning that the Maclaurin sequence and the Newtonian bifurcation sequences as well as the disc limit presented in Section 2.1 are all to be found as the Newtonian limits of the strange matter configurations. If in a generic class containing three Newtonian sequences, the two relativistic adjoining sequences meet, then the class structure here will be identical to that for homogeneous matter. That is not to say, however, that the properties of the member constituents need be similar in a highly relativistic regime.

Figure 3.33 represents the counterpart to Fig. 3.17 and shows the boundary curves for the first class of solutions as well as the entire Maclaurin sequence. The normalized square of the angular velocity $\Omega^2/16\pi B$ is plotted here versus the radius ratio A and is not divided by the central density as in Figs 3.32 and 3.30. Thus this plot clearly reflects the fact that Ω does not reach its maximum along the sequence of infinite pressure. Its maximal value is to be found for a central pressure $p_c/B =: \tilde{p}_c \approx 64$ along the mass-shedding curve. Curves of constant central pressure approach the limiting sequence $\tilde{p}_c = \infty$ in an oscillatory manner

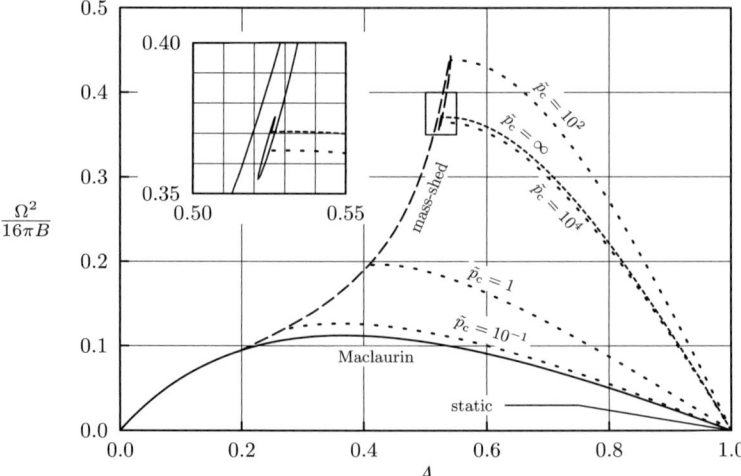

Fig. 3.33. The dependence of $\Omega^2/16\pi B$ on A is shown for the boundary curves of the first class of strange matter configurations. In addition, various sequences with constant central pressure $\tilde{p}_c := p_c/B$ are plotted using dotted lines. Line types are the same as those used in Fig. 3.14, with the exception of the mass-shedding curve in the inset, which is drawn using a solid line to render greater detail.

as shown vividly by the blowup. As with other equations of state considered here, further physical parameters such as angular momentum, mass, baryonic mass and redshift oscillate along the mass-shedding curve. An example of the typical inspiral behaviour along this sequence is provided in Fig. 3.34 for redshift z. The maximal values are listed in Table 3.5. It is notable that the increase in the maximal mass due to rotation is 44% and thus significantly greater than for any other equation of state considered in this book, including homogeneous matter.

The second class of solutions naturally possesses the same three Newtonian limiting sequences as with homogeneous matter. The pertinent piece of the Maclaurin sequence joins on to two bifurcating sequences ending in a mass-shedding limit in the one case and an infinitely thin ring in the second (Fischer *et al.* 2005, Labranche *et al.* 2007). Emanating from these two points are the mass-shedding and extreme Kerr sequences, both of which meet at a point, thus enclosing the class. But for the fact that the three Newtonian curves merge to a single one for polytropes and for a completely degenerate ideal gas of neutrons, these qualitative features are shared by all the equations of state considered in this book. In particular, the existence of a transition to a black hole is a generic feature. A plot of the mass-shedding sequence running from the Newtonian to the black hole limit can be found for a variety of equations of state in Fig. 3.35. One can see that, at least in the parameter pair (V_0, A), the mass-shedding sequence for homogeneous rings

3.4 Configurations with other equations of state

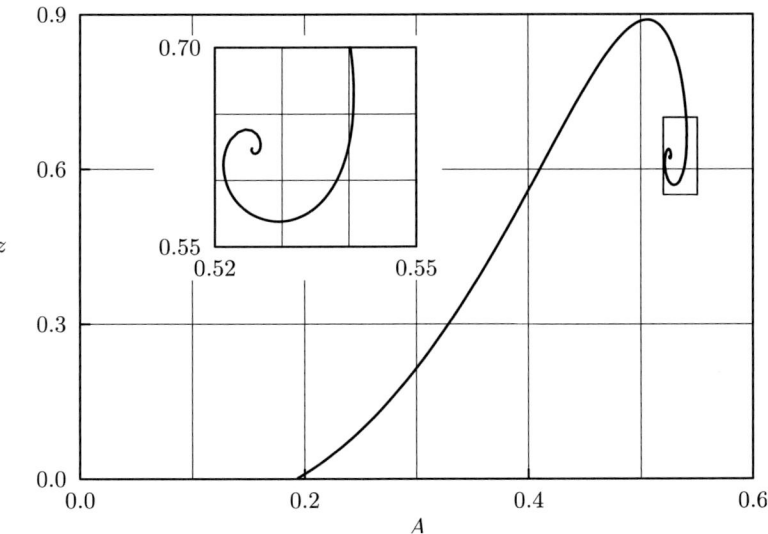

Fig. 3.34. The inspiral behaviour of the redshift z versus the radius ratio A can be seen for the mass-shedding sequence of strange matter configurations.

Table 3.5. *Maximal values for physical quantities are provided for strange matter stars, where 'maximal' refers here to a global maximum for Class I.*

Physical quantity	Value	A
Gravitational mass	$M_{\text{static}} = 0.0258 \; B^{-1/2}$	1.000
Gravitational mass	$M = 0.0372 \; B^{-1/2}$	0.468
Baryonic mass	$h(0)M_0 = 0.0444 \; B^{-1/2}$	0.471
Angular velocity	$\Omega = 4.72 \; B^{1/2}$	0.540
Angular momentum	$J = 0.00123 \; B^{-1}$	0.456
Redshift	$z = 0.890$	0.505

differs only very slightly from that for strange matter rings. Such rings, as well as polytropes with sufficiently small polytropic index n, exhibit a change of topologies. The mass-shedding curve indicates a boundary for A at a given V_0 so that only the shaded region to the left of it is accessible. For every equation of state plotted here there exists a shaded region near the black hole limit. Rings approaching this limit can be considered to be sources of the extreme Kerr metric. The metric generated by them becomes arbitrarily close to that of the extreme Kerr black hole exterior to its horizon.

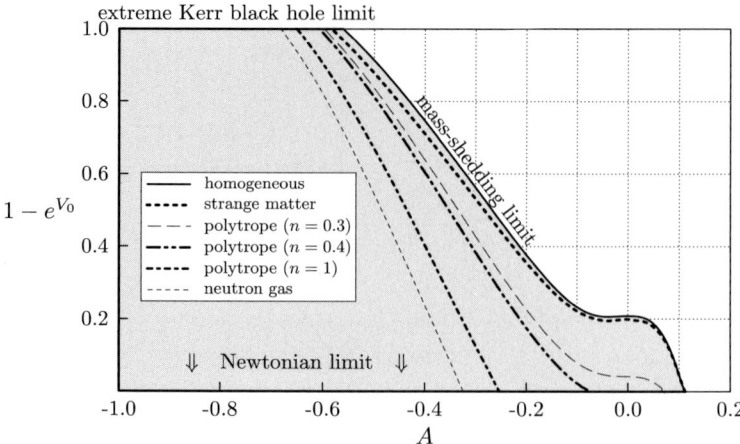

Fig. 3.35. Mass-shedding sequences from the second class for various equations of state are shown. All of them connect the Newtonian with the extreme Kerr black hole limit (after Fischer *et al.* 2005).

3.5 Fluid rings with a central black hole

The figures of equilibrium that have occupied our attention until now have been single objects, namely discs, stars, rings or black holes. Through the detailed exploration of the solution space for homogeneous bodies, we were led naturally to the verge of a two-body problem, for example a star with a surrounding ring in Subsection 3.3.4, and chose this to mark a boundary in solution space. Although two-body systems are not what one first thinks of upon hearing 'figures of equilibrium', they can be of considerable interest and will be considered here with a furtive glance toward more distant horizons. It is fairly certain that systems in equilibrium must be axially symmetric (see Section 1.2). In Newtonian theory, spheroidal bodies in axisymmetry cannot be stationary when aligned along the axis of symmetry. In general relativity, all attempts to find multiple (uncharged) bodies aligned along the axis and kept in equilibrium by gravitomagnetic effects have failed. When considering stationary, two-body systems, one is thus led quite naturally to a central figure surrounded by a ring.

In the remainder of this chapter we study such systems. The central object considered here will either be a Newtonian point mass, a black hole or an infinitely flattened, rigidly rotating disc of dust. Spacetimes with central black holes are of interest concerning the study of (i) the collapse of a single neutron star to a black hole and (ii) the coalescence of two compact objects, for it is expected that such systems exist, if only for a short time, see e.g. Shibata *et al.* (2003). In addition to astrophysical motivations, there is also interest in studying a black-hole–ring system in order to see how matter affects the properties of the black hole.

3.5 Fluid rings with a central black hole

Studies of systems consisting of a central rotating black hole and surrounding matter have been carried out by various authors (Will 1974, 1975, Abramowicz *et al.* 1983, 1984, Bodo and Curir 1992, Rezzolla *et al.* 2003, Montero *et al.* 2004, Zanotti *et al.* 2005, Lanza 1992). Using a formulation of Einstein's equations in terms of integral equations, Nishida and Eriguchi (1994) numerically solved the problem of a differentially rotating polytropic perfect fluid ring surrounding a black hole. The methods used (about fifteen years ago now) did not allow for an accuracy high enough to resolve the impact of the matter distribution on the black hole completely, and the authors were misled into making incorrect conjectures regarding the shape of black holes with zero angular momentum, see Ansorg and Petroff (2005).

Horizon and disc boundary conditions

We begin by revisiting the horizon boundary conditions (1.129) and adding corresponding expressions for the remaining metric potentials. For disc–ring systems, a similar introduction of the disc boundary conditions will be provided.

The boundary conditions that hold for a stationary, axisymmetric, asymptotically flat spacetime containing a black hole and a fluid with purely rotational motions were discussed lucidly and at length in Bardeen (1973a). Let, in our coordinates, the central black hole's horizon be described by a constant radius $r = r_h$, with spherical coordinates (r, ϑ) introduced through $\varrho = r \sin \vartheta, \zeta = r \cos \vartheta$. By virtue of the field equations (3.2), the regularity requirements of the metric functions imply the following boundary conditions valid at the horizon $r = r_h$:

$$W = Br_h \sin \vartheta = 0 \quad \Rightarrow \quad B = 0, \tag{3.59}$$

$$\omega = \Omega_h = \text{constant}, \tag{3.60}$$

$$\frac{\partial u}{\partial r} = \frac{1}{r_h}, \tag{3.61}$$

$$\frac{\partial \alpha}{\partial r} = -\frac{1}{r_h}. \tag{3.62}$$

Note that in contrast to v, the auxiliary metric function $u = v - \ln B$ remains regular at the horizon.

It is furthermore possible to introduce an additional constant κ defined through the following horizon values:

$$\kappa = e^{-\alpha} \frac{\partial}{\partial r} e^v \bigg|_{r=r_h}. \tag{3.63}$$

This constant is called the surface gravity of the horizon and describes a rescaled gravitational acceleration of a zero angular momentum observer on the horizon, see

Bardeen (1973a). We characterize a degenerate non-vanishing black hole (a black hole with degenerate horizon) by $\kappa = 0$ with finite horizon area A.

For a system consisting of a rigidly rotating central disc of dust and a surrounding ring of matter, the conditions for the metric, valid within the disc, are given in Section 1.7.3. In terms of the coefficients (α, B, ω, ν) they read:

$$e^{2V} = e^{2\nu}(1-v^2) = e^{2V_d} = \text{constant}, \tag{3.64}$$

$$B_{,\zeta} = 0, \tag{3.65}$$

$$4(\Omega_d - \omega)v_{,\zeta} + (1+v^2)\omega_{,\zeta} = 0, \tag{3.66}$$

$$(1-v^2)\nu_{,\zeta} + (1+v^2)\alpha_{,\zeta} = 0. \tag{3.67}$$

Here Ω_d is the constant angular velocity of the disc and V_d the constant value of the function V defined in (1.19). This value is related to the relative redshift of zero angular momentum photons emitted from the disc's surface and received at infinity, see (1.28).

A Newtonian ring surrounding a point mass

Before turning our attention to a black hole surrounded by a ring, it is natural to study a related Newtonian problem first. As discussed in Ansorg and Petroff (2005), a Newtonian, rigidly rotating test-ring of finite size cannot exist in equilibrium. Since we know that rings without a central body exist within Newtonian theory (Kowalewsky 1885, Poincaré 1885, Dyson 1892, 1893, Wong 1974, Eriguchi and Sugimoto 1981, Ansorg et al. 2003c), there must exist a maximum for the ratio of the mass of the central body to that of the ring if the ring is to remain a finite size. One would expect the gravitational pull towards the central object to grow ever stronger, until mass-shedding at the inner edge sets in, i.e. until the gradient of pressure at the inner edge of the ring in the equatorial plane vanishes and a cusp develops, marking the point at which a fluid element is about to be pulled away from the ring. This is indeed what is observed. In Fig. 3.36, a sequence of rings about a point mass is shown for an increasing ratio of the central to the ring mass M_c/M_r. The ratio of the inner to outer radius of the ring was held constant at the value $A = -0.6$ and the total (normalized) mass of the system was taken to be $M_{\text{tot}}\sqrt{\mu} = 1$.

If we consider the sequence of configurations at the inner mass-shedding limit and with constant total mass, then we can vary a third parameter such as M_c/M_r. In the limit for which this ratio of masses goes to zero (i.e. when the point mass vanishes), we arrive at the configuration denoted by '∗' in Fig. 3.19. As described there, such a configuration can be found along the sequence bifurcating from the

3.5 Fluid rings with a central black hole

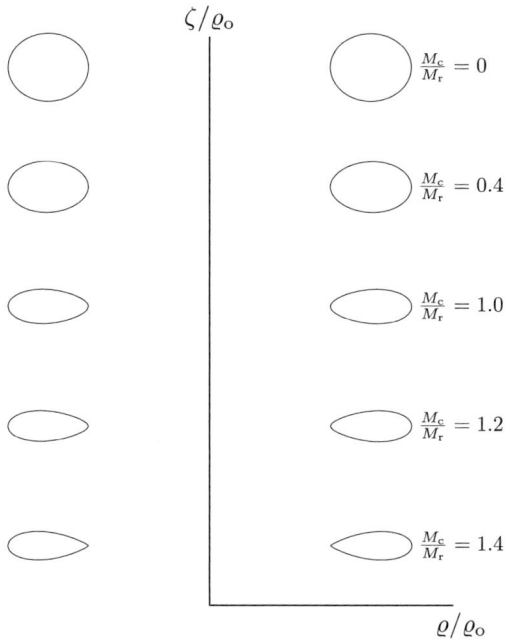

Fig. 3.36. Cross-sections of Newtonian rings surrounding a point mass with varying ratios of central to ring mass M_c/M_r. The normalized coordinate ζ/ϱ_o is plotted against ϱ/ϱ_o. For each of these configurations, the ratio of inner to outer radius of the ring was chosen to be $A = -0.6$ and for the normalized total mass we took $M_{\text{tot}}\sqrt{\mu} = (M_c + M_r)\sqrt{\mu} = 1$ (after Ansorg and Petroff 2005).

Maclaurin spheroid with an eccentricity of $\epsilon = 0.98523\ldots$ and marks the transition from a spheroidal to a toroidal topology. Presumably, there is no upper limit to the value of M_c/M_r that can be reached. However, this test-ring limit could only be reached if the ring were not of finite size, i.e. in the limit $M_c/M_r \to \infty$ it follows that $A \to -1$. The cross-sections for configurations with an inner mass-shed can be seen in Fig. 3.37.

Negative Komar mass of central objects

For stationary black-hole–ring and disc–ring systems, it is possible to assign a Komar mass and angular momentum to each of the two objects, see Komar (1959) and Bardeen (1973a). In particular, these expressions are obtained by considering the integrals (1.57) over a spacelike volume $\mathcal{V} \subset \Sigma$ containing only the object in question. By virtue of the field equations (3.2), these volume integrals can be rewritten in terms of surface integrals over the boundary $\partial \mathcal{V}$. In this manner it is also possible to assign the Komar quantities to a black hole. For a spacetime containing only a single object, this definition of the Komar mass coincides with that of the

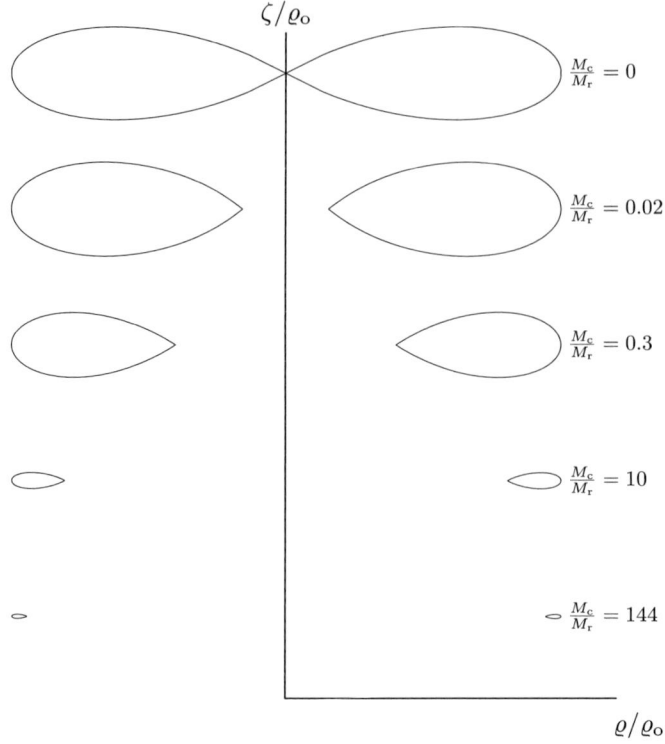

Fig. 3.37. Cross-sections of Newtonian rings surrounding a point mass with varying ratios of central to ring mass M_c/M_r. The normalized coordinate ζ/ϱ_0 is plotted against ϱ/ϱ_0. Each of these configurations possesses an inner mass-shed and has a normalized total mass of $M_{\text{tot}}\sqrt{\epsilon} = (M_c + M_r)\sqrt{\epsilon} = 1$ (after Ansorg and Petroff 2005).

gravitational mass. For a system with two or more objects, the gravitational mass is equal to the total Komar mass, which is the sum of the individual Komar masses.[6]

If one deals with the Komar mass, a natural question concerns its positive definiteness. We address this question by analysing the following formulae

$$M_h = \frac{\kappa A}{4\pi} + 2\Omega_h J_h \qquad (3.68)$$

and

$$M_d = e^{V_d} M_0 + 2\Omega_d J_d, \qquad (3.69)$$

valid for central black hole and central disc configurations respectively.

[6] Note that some authors use the term Komar mass exclusively for what we call here the total Komar mass. The total Komar mass must, of course, obey the positive mass theorem (Schoen and Yau 1979).

3.5 Fluid rings with a central black hole

In Equation (3.68), the central black hole's Komar mass M_h is related to (i) its surface gravity κ, (ii) its horizon area A, (iii) the angular velocity Ω_h of the horizon, and (iv) the black hole's (Komar) angular momentum J_h. For single black holes, Equation (3.68) was given by Smarr (1973), but it holds true even in the presence of a surrounding ring (Bardeen 1973a, see also Carter 1973).

A similar expression can be derived for rigidly rotating discs of dust with and without a surrounding ring of matter, cf. (2.237). In Equation (3.69) the disc's Komar mass M_d is given in terms of (i) the constant e^{V_d}, (ii) the baryonic mass M_0 of the central disc, (iii) its angular velocity Ω_d, and (iv) its (Komar) angular momentum J_d.

The first summands on the right hand sides of formulae (3.68) and (3.69) are always positive. However, each of these terms can become small if we assume the horizon area and the baryonic mass to be finite and consider the central object to be close to a degenerate black hole. As will be discussed below, we find a continuous transition from the central disc to the central black hole configurations (Ansorg and Petroff 2006), and at the transition point, the central object is a degenerate black hole for which the first terms in equations (3.68, 3.69) vanish.

For the discussion of the sign of the second summands, a 'frame-dragging' effect of the central object caused by the surrounding ring is important. If the torus is highly relativistic and quickly rotating, it creates a large ergosphere, see Subsection 1.6.2. In this case, a counter-rotating central object (i.e. the sign of its angular momentum is opposite to that of the torus) inside the ergosphere is dragged along the direction of the motion of the ring's fluid elements. As a consequence, the corresponding *angular velocity* of the central object can assume the same sign as that of the surrounding ring, and thus the second summand becomes negative.

Combining the two arguments, it is possible to identify negative Komar masses by considering central objects close to a degenerate black hole and counter-rotating with respect to the torus. Note that only highly relativistic and quickly rotating tori will exert a sufficiently large frame-dragging effect to bring this about. Moreover, the specific rate of counter-rotation must be limited since very strong counter-rotation would lead to opposite signs of the two angular velocities, $\Omega_{h/d}$ and Ω_{ring}, and hence to a positive second summand.

In Fig. 3.38 we display sequences of both central black hole and central disc configurations along which the Komar mass of the central object becomes negative. Along the sequence the ratio of the circumferential inner radius $R_i^{circ} = \sqrt{g_{\varphi\varphi}(\varrho = \varrho_i, \zeta = 0)}$ to that of the outer one $R_o^{circ} = \sqrt{g_{\varphi\varphi}(\varrho = \varrho_o, \zeta = 0)}$ was held constant at the value $R_i^{circ}/R_o^{circ} = 0.85$. Furthermore, the outer mass-shed parameter from Equation (3.51) was chosen to be $\beta = 0.3$ and as a third parameter $\varrho_d/\varrho_o = 0.1$ was chosen for the disc or $r_h/\varrho_o = 0.1$ for the black hole.

172 *Numerical treatment of the general case*

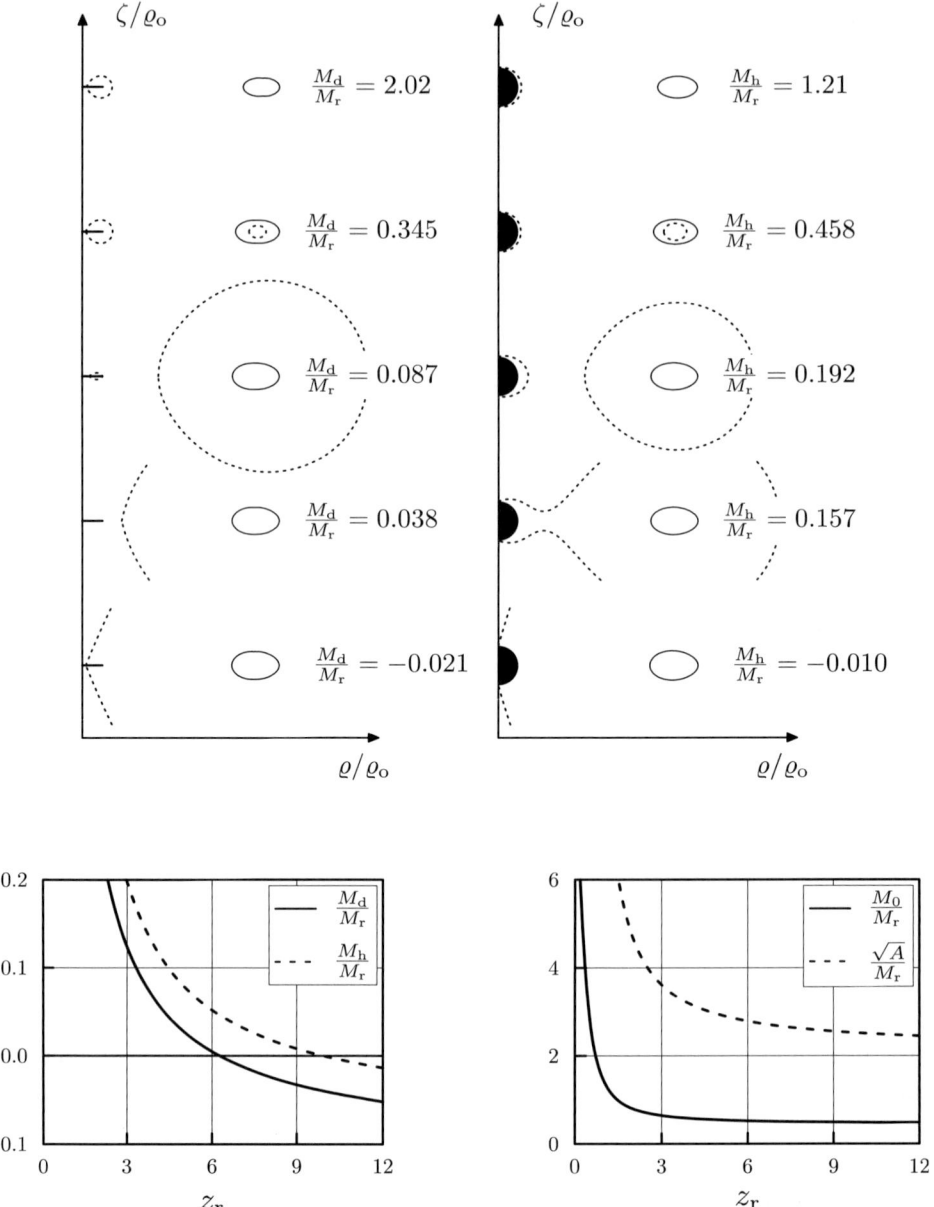

Fig. 3.38. On the lower left, the ratio of the Komar mass of the central object to that of the ring is plotted versus the ring's redshift z_r for a sequence with $R_i^{\mathrm{circ}}/R_o^{\mathrm{circ}} = 0.85$, $\beta = 0.3$ and $\varrho_d/\varrho_o = 0.1$ for the disc or $r_h/\varrho_o = 0.1$ for the black hole (see text for an explanation of the symbols). Knowing that M_r remains positive, one can see that $M_{d/h}$ becomes negative. The lower right shows a similar plot, but containing the disc's baryonic mass and the square root of the horizon area. Above the plots, the coordinate shape of the ring and central object (solid lines) and their ergospheres (dotted lines) are shown for these sequences (after Ansorg and Petroff 2006).

3.5 Fluid rings with a central black hole

The evolution of the coordinate shape of the ring, the central object and their ergospheres can be followed in Fig. 3.38. Looking first at the series of pictures on the left, we begin with a fairly 'Newtonian' ring and can see that only the disc possesses an ergosphere. As explained above, the ring and the disc must be counter-rotating, i.e. their angular momenta have opposite signs. For weakly relativistic rings with a small influence on the central object, the signs of the central angular momentum and velocity coincide, resulting in a positive Komar mass for the central object. However, as the ring becomes increasingly relativistic and develops an ergosphere, its frame-dragging effect on the disc becomes more pronounced, causing the disc's angular velocity to decrease, whence its ergosphere shrinks and finally vanishes. Going even further, the frame-dragging finally forces the disc to corotate with the ring although its angular momentum still has the opposite sign. Relative to the size of the ring, the ergosphere grows very large, which is why we show only a portion of its boundary in the last two pictures in the sequence. From an outside observer's perspective, the configuration is shrinking toward the centre and the outside metric is beginning to resemble that of the extreme Kerr metric. The ring's ergosphere continues to grow, finally engulfing the disc. After a good portion of the disc finds itself inside the ergosphere, the frame dragging becomes significant enough that the magnitude of $\Omega_d J_d$ is sufficiently large to result in a negative mass. The series of pictures on the right is the counterpart for a black hole and shows similar behaviour to the disc case. A black hole with non-vanishing Ω_h always has an ergosphere surrounding it however. Its sense of rotation must agree with that of the ring before their ergospheres can merge, independent of the sign of J_h.

We now devote our attention to the behaviour of the Komar mass in the parametric transition from the central disc of dust to a black hole. A generalization of the proof given in Section 1.8.3 (see also Meinel 2004, 2006) implies that such a transition exists if and only if V_d tends to $-\infty$, which in turn implies that $M_d = 2\Omega_d J_d$ must hold.[7] The equality of (3.69) and (3.68) then requires for non-vanishing black holes that $\kappa = 0$. The plot on the lower left of Fig. 3.39 shows that such transitions do indeed exist. Here the Komar mass ratio was chosen as an exemplary parameter and plotted versus a measure of the distance to the transition point representing a degenerate black hole surrounded by a ring. The plot on the lower right suggests a very similar transition to the one known analytically for the rigidly rotating disc of dust without a ring as can be seen by comparing it to Fig. 2.17. The picture sequence in Fig. 3.39 shows the evolution of the coordinate shapes of these configurations.

The results presented reveal clearly that the Komar mass is not an intrinsic property of a gravitational source but rather a feature of an object within a highly

---

[7] The generalization of the arguments in Meinel (2006) requires the assumption that $-\xi^i u_i$ as defined there is bounded from below. In the presence of a surrounding ring, $\eta^i u_i$ need not have the same sign as Ω_d.

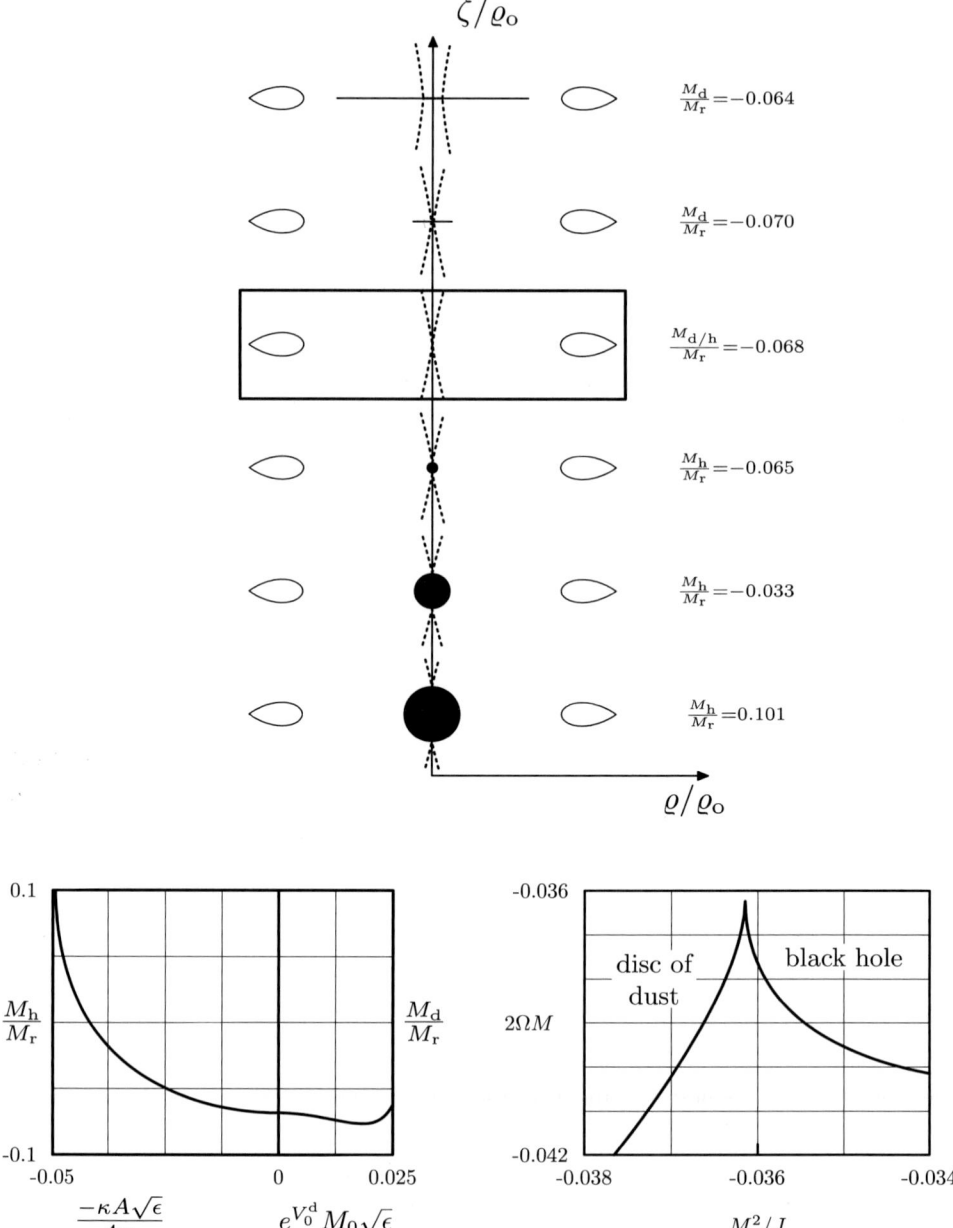

Fig. 3.39. On the lower left, the ratio of the Komar mass of the central object to that of the ring is plotted versus a measure of the distance to the degenerate black hole solution. The sequences are defined by $R_i^{\text{circ}}/R_o^{\text{circ}} = 0.85$, $\beta = 0$ for the outer mass-shed parameter and $V_0^r = -2.7$ for the ring's corotating surface potential. On the lower right, a plot similar to Fig. 2.17 is shown for the transition. Up above, the coordinate shape of the ring and central object (solid lines) and their ergospheres (dotted lines) are drawn. The framed picture indicates the transition point from the disc to the black hole (after Ansorg and Petroff 2006).

relativistic spacetime geometry. The discrepancy between the sign of $\Omega_{h/d}$ and $J_{h/d}$, which is responsible for the negative Komar masses, is a result of the fact that the central body's motion is determined by the local environment, whereas the angular velocity refers to rotation 'with respect to infinity' (see Section 1.1). Since the individual Komar mass is not a good measure of the intrinsic attributes of a local object, it is natural to consider other local mass definitions. An interesting candidate for such a definition in the case of black holes is the Christodoulou mass (Christodoulou 1970), which will be discussed shortly.

Black holes with degenerate horizon

It has been shown by Ansorg and Pfister (2008) that the horizon area A_h and the angular momentum J_h of a degenerate black hole ($\kappa = 0$) are related by

$$8\pi |J_h| = A_h, \tag{3.70}$$

even in the presence of a surrounding fluid ring. The proof runs as follows:

First, a new radial coordinate R is introduced

$$R = \frac{1}{2}\left(r + \frac{r_h^2}{r}\right), \tag{3.71}$$

in which, as stated by Bardeen (1973a), pages 251–252, the following functions of the metric potentials are positive and regular with respect to R and $\cos\vartheta$ in the vicinity of the black hole, even when the degenerate limit is encountered:[8]

$$\hat{\mu} = r^2 e^{2\mu}, \tag{3.72}$$

$$\hat{u} = r^2 e^{-2u}, \tag{3.73}$$

$$\hat{B} = \frac{r}{\sqrt{R^2 - r_h^2}} B. \tag{3.74}$$

The coordinate R penetrates the horizon, that is, spatial points with coordinate values $R < r_h$ are inside the horizon. Note that, in contrast, for any value $r > 0$ we obtain $R \geq r_h$, i.e. the coordinate r is not horizon penetrating.

In terms of these functions, the surface gravity can be expressed as:

$$\kappa = r_h \hat{B} \left(\hat{\mu}\hat{u}\right)^{-1/2} = \text{constant}.$$

Hence a degenerate black hole with $\kappa = 0$ is characterized by $r_h = 0$.

It can now be shown that the Einstein equations, written for the degenerate limit in terms of the above regular functions in (R, ϑ) and evaluated at the horizon

---

[8] The quantities introduced here are closely related to Bardeen's expressions: $h = 2r_h, \lambda = 2R, B_R = \hat{B}/2$.

$R = r_h = 0$, do not contain terms in which R-derivatives are involved. As a consequence, it becomes possible to work out the horizon boundary values of all metric functions explicitly. If these expressions are inserted into the integral formulae for the black hole angular momentum and horizon area, one obtains (3.70), which is valid for axially and equatorially symmetric, stationary configurations consisting of a degenerate central black hole with surrounding matter. For more details of the proof, see Ansorg and Pfister (2008). Moreover, overwhelming evidence from numerical calculations is presented there to support the conjecture that

$$8\pi |J_h| \leq A_h \tag{3.75}$$

holds true for any axially and equatorially symmetric, stationary configuration consisting of a central black hole with surrounding perfect fluid matter, and that, in particular, the equals sign holds if and only if the central black hole is degenerate.

A well-known relation for degenerate Kerr black holes is given by

$$M^2 = |J|, \tag{3.76}$$

see Subsection 1.8.2. However, this relation is no longer true if M is taken to be the Komar mass M_h and we allow for matter surrounding the black hole. In fact, we have seen that the Komar mass can vanish for a rotating degenerate black hole with finite angular momentum J_h.

It is interesting to note that (3.76) does hold however, by virtue of (3.70), even when the degenerate black hole is surrounded by additional matter, if one takes M to be the Christodoulou mass

$$M_C = \sqrt{\frac{A_h}{16\pi} + \frac{4\pi J_h^2}{A_h}}, \tag{3.77}$$

which plays a fundamental role in the isolated and dynamical horizon formalism, see Ashtekar and Krishnan (2004) for an overview. Moreover, this mass parameter is being used widely in the field of dynamical calculations of spacetimes containing black holes, e.g. for describing black holes in the centre of accretion discs (Font 2003, and references therein).

4
Remarks on stability and astrophysical relevance

In general, a relativistic figure of equilibrium as calculated in the previous chapters is to be expected to exist in nature only if it is in *stable* equilibrium. Therefore, in addition to other aspects, like realistic equations of state, magnetic fields and initial conditions, the investigation of stability properties is very important for identifying configurations that might be astrophysically relevant. A complete stability analysis of relativistic figures of equilibrium is extremely difficult. Moreover, the stability depends on matter properties like viscosity and thermal conductivity, which are unimportant for the equilibrium state itself and therefore do not need to be specified in this book. Our intention, as expressed in the preface, is to 'place emphasis on the rigorous treatment of simple models instead of trying to describe real objects with their many complex facets' and, consequently, an extensive treatment of stability questions is beyond the scope of this book. Nevertheless, in the following we will discuss some aspects of stability of rotating fluid configurations in general relativity.

Stability with respect to axisymmetric perturbations

Friedman *et al.* (1988) have shown that a version of the turning-point method going back to Poincaré (1885), who investigated the stability of Newtonian equilibrium configurations (cf. Subsection 3.3.1), can be used to locate points along sequences of relativistic figures of equilibrium at which secular instability[1] with respect to axisymmetric perturbations sets in, see also Thorne (1967). The result can be formulated as follows: Whenever a continuous sequence of equilibrium configurations labelled by a parameter λ, with a given equation of state $\epsilon = \epsilon(p)$ and constant angular momentum J, has an extremum of the gravitational mass M at a point $\lambda = \lambda_0$, this point marks the onset of instability. The unstable part of the

---

[1] A 'secular' instability – in contrast to a 'dynamical' one – requires some dissipative mechanism, provided for example by viscosity, in order to take effect.

sequence, near λ_0, can be identified by the condition

$$\frac{\mathrm{d}^2 M}{\mathrm{d}M_0^2} > 0 \tag{4.1}$$

or, equivalently,

$$\frac{\mathrm{d}V_0}{\mathrm{d}\lambda} \frac{\mathrm{d}M}{\mathrm{d}\lambda} > 0. \tag{4.2}$$

(It is assumed that $\mathrm{d}[(\mathrm{d}V_0/\mathrm{d}\lambda)(\mathrm{d}M/\mathrm{d}\lambda)]/\mathrm{d}\lambda \neq 0$ at $\lambda = \lambda_0$.) This result can easily be understood by means of the relation (1.59) between variations of M, J and the baryonic mass M_0, specialized to a continuous sequence of equilibrium configurations, where all quantities depend on the single parameter λ,

$$\frac{\mathrm{d}M}{\mathrm{d}\lambda} = \Omega \frac{\mathrm{d}J}{\mathrm{d}\lambda} + h(0)\mathrm{e}^{V_0} \frac{\mathrm{d}M_0}{\mathrm{d}\lambda}. \tag{4.3}$$

For a sequence of constant angular momentum J, an extremum in the gravitational mass M at some point $\lambda = \lambda_0$ automatically results in an extremum in the baryonic mass M_0 at the same point. This means that there exist two nearby points λ_1, λ_2 of the sequence ($\lambda_1 < \lambda_0 < \lambda_2$) for some value of M_0 near its extremum. The configurations at these two points thus have the same M_0 and J, but they differ in M. Provided that a suitable dissipative mechanism is available, the configuration with larger M is unstable: it can evolve towards the other one as a result of small perturbations. Note that equilibrium figures are configurations, which, for prescribed M_0 and J, extremize M in a well-defined sense (Hartle and Sharp 1967). Stable equilibria must, of course, be minima of M. An axisymmetric time evolution does not change the total angular momentum J, but viscosity can lead to a redistribution of angular momentum (transfer from one baryonic fluid ring to another). This process is essential for reaching an equilibrium state with its uniform rotation.

The equivalence of (4.1) and (4.2) follows immediately from (4.3) and $J =$ constant:

$$\frac{\mathrm{d}M}{\mathrm{d}M_0} = h(0)\mathrm{e}^{V_0} \Rightarrow \frac{\mathrm{d}^2 M}{\mathrm{d}M_0^2} = h(0)\mathrm{e}^{V_0} \frac{\mathrm{d}V_0}{\mathrm{d}\lambda} \frac{\mathrm{d}\lambda}{\mathrm{d}M_0}$$

$$= \frac{\mathrm{d}V_0}{\mathrm{d}\lambda} \frac{\mathrm{d}M}{\mathrm{d}\lambda} \left(\frac{\mathrm{d}M_0}{\mathrm{d}\lambda}\right)^{-2}. \tag{4.4}$$

This relation also shows that the second derivative $\mathrm{d}^2 M/\mathrm{d}M_0^2$ diverges at $\lambda = \lambda_0$ and changes its sign, leading to a cusp in an M_0-M diagram, see Fig. 4.1. Independent of the question of whether the turning point of M (and M_0) is a maximum or a minimum, the larger M-value for given M_0 will be found on that

Remarks on stability and astrophysical relevance 179

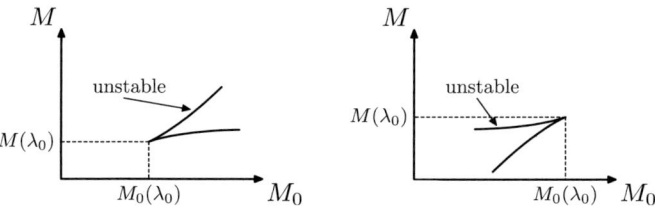

Fig. 4.1. Schematic M_0-M diagram of a J = constant sequence of equilibrium figures with a given equation of state near a minimum (left picture) or a maximum (right picture) of M and M_0.

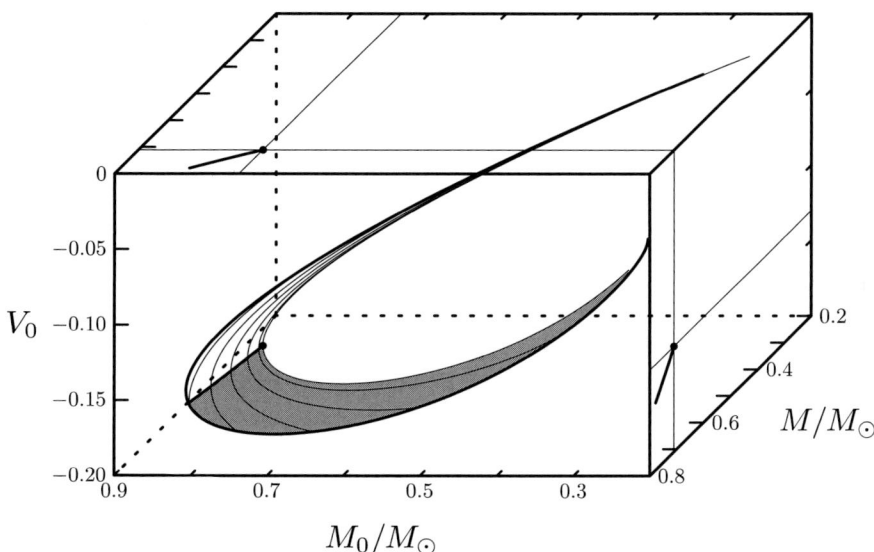

Fig. 4.2. Partial sequences of constant angular momentum J for stars governed by the equation of state for a completely degenerate, ideal gas of neutrons. Moving from the inside outward, the values of $J/K_n^{3/2}$ are 0, 0.006, 0.012, 0.018, 0.024 and 0.03. For the meaning of the normalization constant K_n see (3.57). The long thicker line at the outside is a portion of the mass-shedding curve. The short thicker line marks the position of maximal mass for each sequence and is thus the boundary of the secularly unstable region, which is here marked by grey shading. One end of this boundary line, marked by a dot, corresponds to the $J = 0$ configuration of maximal mass. The projections of this line onto two planes in the $M_0 - M - V_0$ space are also shown.

part of the sequence which satisfies (4.1). Note that the instability condition can also be written as

$$\frac{dM}{dV_0} > 0. \qquad (4.5)$$

For $dV_0/d\lambda \neq 0$ at $\lambda = \lambda_0$, the parameter V_0 can be used itself to label the equilibrium sequence near the turning point. If one determines the turning points of such $J = $ constant sequences in dependence of J, a curve in the two-dimensional manifold of solutions (equilibrium figures for the given equation of state) is obtained, which represents the boundary of the secularly unstable region.[2] An example is given in Fig. 4.2.

A sufficient condition for dynamical stability with respect to small axisymmetric perturbations is the positive-definiteness of a well-defined energy functional E for all relevant perturbations (Chandrasekhar and Friedman 1972a,b, 1973, Schutz 1972), see also Friedman and Ipser (1992) – since E can only decrease as a consequence of (outgoing) gravitational radiation. (If positive definite, E plays the role of a Lyapunov functional.) A more practicable method to investigate dynamical stability can be the direct numerical simulation of the time evolution of slightly perturbed equilibrium configurations with a purely axisymmetric code. However, in general, a sequence of equilibrium figures becomes dynamically unstable after it has become secularly unstable, see Friedman et al. (1988). Therefore, the simple turning-point criterion discussed above may be sufficient to find the limits of axisymmetric stability.

Stability with respect to non-axisymmetric perturbations

A complete stability analysis must, of course, also include the consideration of non-axisymmetric perturbations. A remarkable result is that all rotating perfect-fluid equilibrium configurations are unstable to non-axisymmetric perturbations by a mechanism that leads to a loss of angular momentum via gravitational radiation. This 'Chandrasekhar–Friedman–Schutz (CFS) instability' (Chandrasekhar 1970, Friedman and Schutz 1978) can, however, be damped by viscosity, see Detweiler and Lindblom (1977) and Lindblom and Hiscock (1983).

Many more results on axisymmetric as well as non-axisymmetric stability of relativistic stars can be found in Stergioulas (2003) and references therein.

---

[2] Note that the same curve can also be obtained by considering turning points of M (and J) for sequences $M_0 = $ constant.

Appendix 1
A detailed look at the mass-shedding limit

A1.1 The differentiability of functions at the mass-shedding limit

Since solutions to Poisson-like equations on domains containing corners are known to be non-analytic in general, it will come as no surprise to learn that the functions involved in describing a mass-shedding configuration are also not analytic. It is of interest, especially for the numerical methods used in this book, to be more precise. Ideally, we would be able to determine the asymptotic behaviour of the functions involved as one approaches a corner. We shall content ourselves in a first analysis, however, with determining which derivatives become singular.

The behaviour of solutions to Poisson-like equations on domains containing corners has been studied by Wigley (1970), Eisenstat (1974) and discussed in Birkhoff and Lynch (1984). In those analyses, it was always assumed that the domain boundaries are known whereas in our case, the (free) boundary arises from solving a global problem. Although we know that the cross-section of the surface of a mass-shedding body contains a corner, we do not know much, a priori, about the differentiability of its parametric representation $\zeta_b(\varrho)$.

Nonetheless, in order to study the behaviour of the Newtonian potential U, let us imagine that the global problem has been solved, that the surface is known, and that a cusp (i.e. mass shedding, cf. page 26) exists at the point $\varrho = \varrho_0$. Inside the body, we shall refer to the potential as U_i and outside as U_o and leave off the index when an expression is valid for either potential. We assume that the known functions U_i, U_o and ζ_b are C^2 and shall find that this leads to a contradiction.

Taking the derivative of (1.122) with respect to ϱ along the surface yields

$$\frac{\partial^2 U}{\partial \varrho^2} + 2\frac{\partial^2 U}{\partial \varrho\, \partial \zeta}\frac{d\zeta_b}{d\varrho} + \frac{\partial^2 U}{\partial \zeta^2}\left(\frac{d\zeta_b}{d\varrho}\right)^2 + \frac{\partial U}{\partial \zeta}\frac{d^2\zeta_b}{d\varrho^2} = \Omega^2. \qquad (A1.1)$$

As a result of equatorial symmetry, we know that

$$\left.\frac{\partial^{i+j} U}{\partial \varrho^i \partial \zeta^j}\right|_{\zeta=0} = 0 \qquad \text{for } j \text{ odd}. \qquad (A1.2)$$

Therefore, at the point $\varrho = \varrho_0$, (A1.1) reduces to

$$\frac{\partial^2 U}{\partial \varrho^2} + \frac{\partial^2 U}{\partial \zeta^2}\left(\frac{d\zeta_b}{d\varrho}\right)^2 = \Omega^2, \tag{A1.3}$$

remembering that ζ_b is assumed to be a C^2 function.

Making use of (1.124), the Poisson and Laplace equations at the point $\varrho = \varrho_0$ are

$$\frac{\partial^2 U_i}{\partial \varrho^2} + \frac{\partial^2 U_i}{\partial \zeta^2} = 4\pi\mu - \Omega^2, \tag{A1.4a}$$

$$\frac{\partial^2 U_o}{\partial \varrho^2} + \frac{\partial^2 U_o}{\partial \zeta^2} = -\Omega^2. \tag{A1.4b}$$

The system of equations made up of (A1.3) and (A1.4) leads to

$$\frac{\partial^2 U_i}{\partial \zeta^2} = \frac{2\Omega^2 - 4\pi\mu}{\left(\frac{d\zeta_b}{d\varrho}\right)^2 - 1}, \tag{A1.5a}$$

$$\frac{\partial^2 U_o}{\partial \zeta^2} = \frac{2\Omega^2}{\left(\frac{d\zeta_b}{d\varrho}\right)^2 - 1}. \tag{A1.5b}$$

Discounting the possibility that $\Omega = 0$, since we do not expect mass-shedding to occur for non-rotating stars, we find

$$\left(\frac{d\zeta_b}{d\varrho}\right)^2 \neq 1. \tag{A1.6}$$

The transition conditions at the surface of the star require that the first derivatives of U be continuous. Choosing to look at the derivative of U with respect to ζ on the boundary at a point ϱ_1 not far from ϱ_0, we can write

$$\frac{\partial U(\varrho_1, \zeta_b(\varrho_1))}{\partial \zeta} = \frac{\partial U(\varrho_0, 0)}{\partial \zeta}$$
$$+ \left(\frac{\partial^2 U(\varrho_0, 0)}{\partial \zeta \partial \varrho} + \frac{\partial^2 U(\varrho_0, 0)}{\partial \zeta^2}\frac{d\zeta_b}{d\varrho}\right)(\varrho_1 - \varrho_0) + \cdots$$
$$= \frac{\partial^2 U(\varrho_0, 0)}{\partial \zeta^2}\frac{d\zeta_b}{d\varrho}(\varrho_1 - \varrho_0) + \cdots. \tag{A1.7}$$

Since this equation holds both for U_i and U_o, the first derivative is only continuous if

$$\frac{\partial^2 U_i}{\partial \zeta^2} = \frac{\partial^2 U_o}{\partial \zeta^2} \tag{A1.8}$$

A1.1 The differentiability of functions at the mass-shedding limit

holds on the surface at the point ϱ_0. We see from Equations (A1.5) that this can only be true if $\mu = 0$ at this point. In other words, we have arrived at a contradiction for equations of state such as homogeneous or strange matter that allow for the density to jump discontinuously at the surface. For such configurations, the functions U_i, U_o and ζ_b cannot all be C^2.

For other equations of state, we can derive similar results by assuming higher differentiability of the functions. If, for example, we assume now that U_i, U_o and ζ_b are C^3 functions, and consider equations of state for which $\mu = 0$ at the boundary, then we can proceed to differentiate (1.122) again with respect to ϱ. At the point ϱ_0 we then find

$$\frac{\partial^3 U}{\partial \varrho^3} + 3 \frac{\partial^3 U}{\partial \varrho \, \partial \zeta^2} \left(\frac{d\zeta_b}{d\varrho} \right)^2 + 3 \frac{\partial^2 U}{\partial \zeta^2} \frac{d\zeta_b}{d\varrho} \frac{d^2 \zeta_b}{d\varrho^2} = 0. \tag{A1.9}$$

Differentiating the Laplace and Poisson equations with respect to ϱ and eliminating the first and second derivatives with respect to ϱ using (1.124) and (A1.3) yields

$$\frac{\partial^3 U_i}{\partial \varrho^3} + \frac{\partial^3 U_i}{\partial \varrho \, \partial \zeta^2} - \frac{\partial^2 U_i}{\partial \zeta^2} \frac{d\zeta_b}{d\varrho} \left(\frac{1}{\varrho} \frac{d\zeta_b}{d\varrho} + 3 \frac{d^2 \zeta_b}{d\varrho^2} \right) = 4\pi \frac{\partial \mu}{\partial \varrho} \tag{A1.10a}$$

$$\frac{\partial^3 U_o}{\partial \varrho^3} + \frac{\partial^3 U_o}{\partial \varrho \, \partial \zeta^2} - \frac{\partial^2 U_o}{\partial \zeta^2} \frac{d\zeta_b}{d\varrho} \left(\frac{1}{\varrho} \frac{d\zeta_b}{d\varrho} + 3 \frac{d^2 \zeta_b}{d\varrho^2} \right) = 0. \tag{A1.10b}$$

The solution to the system of equations (A1.9) and (A1.10) implies[1] that

$$\left(\frac{d\zeta_b}{d\varrho} \right)^2 \neq \frac{1}{3} \tag{A1.11}$$

based on the assumption that U_i and U_o are C^3 functions. Taking the next term in the expansion of (A1.7) at the point ϱ_0 and remembering the equality (A1.8), we then find

$$\frac{\partial^3 U_i}{\partial \varrho \, \partial \zeta^2} = \frac{\partial^3 U_o}{\partial \varrho \, \partial \zeta^2}. \tag{A1.12}$$

This cannot be fulfilled along with (A1.10) unless the derivative of μ with respect to ϱ vanishes at the equator. This derivative does not vanish for a polytrope with the polytropic index $n = 1$, for example, meaning that for such stars the potentials and the surface function cannot all be C^3 functions.

---

[1] Strictly speaking, one also has to rule out the possibility that $d\zeta_b/d\varrho = -3r_e d^2 \zeta_b / d\varrho^2$. The first and second derivatives of ζ_b are both expected to be negative however, as can be verified for the concrete example of the Roche model in Equation (A1.18).

The arguments presented above can be extended to C^k functions by taking k-th derivatives and finding contradictions for equations of state for which the $(k-2)$-nd derivative of μ does not vanish at the equator.

A1.2 The mass-shedding limit in the Roche model

If the mass of a star is highly concentrated in its centre, then one can expect that it can be modelled by considering a non-gravitating fluid (shell) in the field of a point mass. In Newtonian theory, models of this sort are known as 'Roche models', based on work done by the French scientist Édouard Roche (1873) (see also Zel'dovich and Novikov 1971, Shapiro and Shibata 2002). As we shall see shortly, the mass-shedding limit is particularly easy to handle in this model.

Taking Equation (1.120) and inserting the potential of a point particle, we find that the surface of a fluid rotating rigidly with the angular velocity Ω is defined by the equation

$$V_0 + \frac{M}{\sqrt{\varrho^2 + \zeta^2}} + \tfrac{1}{2}\Omega^2 \varrho^2 = 0. \tag{A1.13}$$

By considering the point $\varrho = 0$, we can relate the constant V_0 to the polar radius r_p

$$V_0 = -\frac{M}{r_p}. \tag{A1.14}$$

For mass-shedding stars, (1.124) leads to the relation

$$M = r_e^3 \Omega^2, \tag{A1.15}$$

and the surface equation (A1.13) becomes

$$r_p \left(\frac{1}{\sqrt{\varrho^2 + \zeta^2}} + \frac{\varrho^2}{2 r_e^3} \right) = 1. \tag{A1.16}$$

Evaluating this expression at the equator $\zeta = 0$ tells us that for mass-shedding fluids in the Roche model

$$\frac{r_p}{r_e} = \frac{2}{3} \tag{A1.17}$$

holds, independent of the equation of state. With this relationship, the curve describing the fluid's surface can be rewritten as

$$\zeta_b = \frac{\sqrt{4 r_e^2 - \varrho^2} \, (r_e^2 - \varrho^2)}{3 r_e^2 - \varrho^2}, \tag{A1.18}$$

Table A1.1. *Physical parameters of Newtonian polytropic stars with the polytropic index n at the mass-shedding limit. The second column provides a measure of how concentrated the mass is in the centre. The third through fifth columns are to be compared with the corresponding values in the Roche model, see Equations (A1.15), (A1.17) and (A1.19) respectively.*

n	$\dfrac{\mu(\varrho/r_e = 0.1, \zeta = 0)}{\mu_c}$	$\dfrac{M}{r_e^3 \Omega^2}$	$\dfrac{r_p}{r_e}$	$\dfrac{d\zeta_b(\varrho = r_e)}{d\varrho}$
0.5	0.99	0.797	0.4380	−1.433
1.0	0.97	0.906	0.5570	−1.606
1.5	0.94	0.957	0.6151	−1.678
2.0	0.89	0.981	0.6434	−1.709
2.5	0.79	0.992	0.6569	−1.723
3.0	0.63	0.997	0.6630	−1.729
3.5	0.38	0.999	0.6655	−1.731

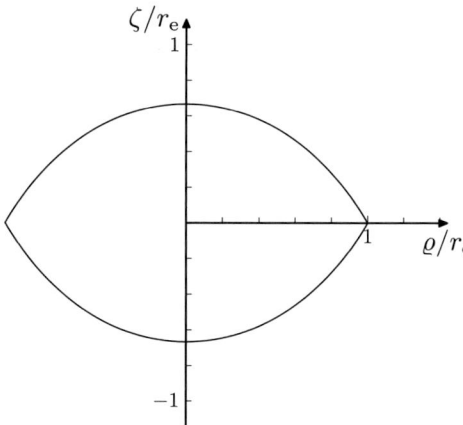

Fig. A1.1. The surface of a mass-shedding star according to the Roche model as given by (A1.18) is depicted in the $\varrho/r_e - \zeta/r_e$ plane.

where the index 'b' was added to conform with the notation in Subsection 1.7.4. It then follows that

$$\left.\frac{d\zeta_b}{d\varrho}\right|_{\zeta=0^+} = -\sqrt{3}, \tag{A1.19}$$

which means the interior angle of the mass-shedding cusp is $2\pi/3$.

The success of the Roche model in describing polytropic stars for large values of the polytropic index n is demonstrated in Table A1.1. The first column in the table indicates the value of n used and the second provides a measure of how highly the mass is concentrated in the star's centre. The remaining three columns demonstrate that the values implied by (A1.15), (A1.17) and (A1.19) are approached for large n, and we see that deviations from the Roche model are on the order of about 0.1% for $n = 3.5$.

Figure A1.1 depicts the shape of the mass-shedding star according to the Roche model [see (A1.18)]. This curve is indistinguishable from that of a polytropic mass-shedding star with $n = 3.5$. The difference between ζ_b/r_e for such a polytropic star and ζ_b/r_e of the Roche model is greatest at the north pole, the value of which can be found in Table A1.1.

Appendix 2
Theta functions: definitions and relations

In this appendix we provide some basic definitions and relations for functions that play an important role in Section 2.3.

We shall not provide the related mathematical theory nor discuss theta functions in generality here. Starting from the (usual) theta functions, which were introduced by Jacobi, we follow the path taken by Rosenhain and define hyperelliptic theta functions of two variables, sometimes called 'ultra-elliptic theta functions' (Rosenhain 1850). Moreover, we list some useful relations between theta functions of each type. Besides the definition of the well-known elliptic integrals of the first and second kinds and the Jacobian elliptic functions, we include two less well-known functions, which can be constructed from elliptic integrals, namely Heuman's lambda function and Jacobi's zeta function. Furthermore, the relations between the Jacobian theta functions and the Jacobian elliptic functions that are important for our purpose in Subsection 2.3.3 are given. Finally, we list derivatives for some of the above functions that were used in Subsection 2.3.4.

Note that the notation in the literature (especially concerning theta functions) is not standardized. Throughout this book we comply strictly with the definitions presented here.

A2.1 Jacobian theta functions

$$\vartheta_1(v;B) := \sum_{n=-\infty}^{\infty} (-1)^n \exp\left\{\left[\tfrac{1}{2}(2n+1)\right]^2 B + (2n+1)v\right\},$$

$$\vartheta_2(v;B) := \sum_{n=-\infty}^{\infty} \exp\left\{\left[\tfrac{1}{2}(2n+1)\right]^2 B + (2n+1)v\right\},$$

$$\vartheta_3(v;B) := \sum_{n=-\infty}^{\infty} \exp\{n^2 B + 2nv\},$$

$$\vartheta_4(v;B) := \sum_{n=-\infty}^{\infty} (-1)^n \exp\{n^2 B + 2nv\}.$$

(A2.1)

Theta functions: definitions and relations

Selected properties of the Jacobian theta functions:

$$\vartheta_1(v+i\pi;B) = -\vartheta_1(v;B), \qquad \vartheta_1\left(v+\tfrac{i\pi}{2};B\right) = i\,\vartheta_2(v;B), \qquad \text{(A2.2)}$$

$$\vartheta_2(v+i\pi;B) = -\vartheta_2(v;B), \qquad \vartheta_2\left(v+\tfrac{i\pi}{2};B\right) = i\,\vartheta_1(v;B), \qquad \text{(A2.3)}$$

$$\vartheta_3(v+i\pi;B) = +\vartheta_3(v;B), \qquad \vartheta_3\left(v+\tfrac{i\pi}{2};B\right) = \vartheta_4(v;B), \qquad \text{(A2.4)}$$

$$\vartheta_4(v+i\pi;B) = +\vartheta_4(v;B), \qquad \vartheta_4\left(v+\tfrac{i\pi}{2};B\right) = \vartheta_3(v;B). \qquad \text{(A2.5)}$$

$$\begin{aligned}\vartheta_1(v+B;B) &= -e^{-(2v+B)}\vartheta_1(v;B),\\ \vartheta_2(v+B;B) &= e^{-(2v+B)}\vartheta_2(v;B),\\ \vartheta_3(v+B;B) &= e^{-(2v+B)}\vartheta_3(v;B),\\ \vartheta_4(v+B;B) &= -e^{-(2v+B)}\vartheta_4(v;B).\end{aligned} \qquad \text{(A2.6)}$$

$$\begin{aligned}\vartheta_1\left(v+\tfrac{B}{2};B\right) &= -e^{-(v+B/4)}\vartheta_4(v;B),\\ \vartheta_2\left(v+\tfrac{B}{2};B\right) &= e^{-(v+B/4)}\vartheta_3(v;B),\\ \vartheta_3\left(v+\tfrac{B}{2};B\right) &= e^{-(v+B/4)}\vartheta_2(v;B),\\ \vartheta_4\left(v+\tfrac{B}{2};B\right) &= e^{-(v+B/4)}\vartheta_1(v;B).\end{aligned} \qquad \text{(A2.7)}$$

It is convenient to introduce the following notation (which is systematic but not common!):

$$\begin{aligned}\Theta_1(u,k) &:= \vartheta_1\left(\frac{\pi u}{2K(k)}, -\pi\frac{K(k')}{K(k)}\right),\\ \Theta_2(u,k) &:= \vartheta_2\left(\frac{\pi u}{2K(k)}, -\pi\frac{K(k')}{K(k)}\right),\\ \Theta_3(u,k) &:= \vartheta_3\left(\frac{\pi u}{2K(k)}, -\pi\frac{K(k')}{K(k)}\right),\\ \Theta_4(u,k) &:= \vartheta_4\left(\frac{\pi u}{2K(k)}, -\pi\frac{K(k')}{K(k)}\right).\end{aligned} \qquad \text{(A2.8)}$$

Special values for $u=0$:

$$\begin{aligned}\Theta_1(0,k) &= \vartheta_1\left(0,-\pi\frac{K(k')}{K(k)}\right) &&= 0,\\ \Theta_2(0,k) &= \vartheta_2\left(0,-\pi\frac{K(k')}{K(k)}\right) &&= \sqrt{\frac{2kK(k)}{\pi}},\end{aligned} \qquad \text{(A2.9)}$$

$$\Theta_3(0,k) = \vartheta_3\left(0, -\pi\frac{K(k')}{K(k)}\right) = \sqrt{\frac{2K(k)}{\pi}},$$

$$\Theta_4(0,k) = \vartheta_4\left(0, -\pi\frac{K(k')}{K(k)}\right) = \sqrt{\frac{2k'K(k)}{\pi}}.$$

A2.2 Rosenhain's theta functions

$\vartheta_{i,k} \equiv \vartheta_{i,k}(v, w; B_{11}, B_{22}, B_{12}), \qquad (i, k \in \{1, 2, 3, 4\})$,

$$\vartheta_{1,k} := \sum_{m=-\infty}^{\infty} (-1)^m \exp\left\{\left[\tfrac{1}{2}(2m+1)\right]^2 B_{11} + (2m+1)v\right\}$$
$$\times \vartheta_k\left(w + \tfrac{1}{2}(2m+1)B_{12}; B_{22}\right),$$

$$\vartheta_{2,k} := \sum_{m=-\infty}^{\infty} \exp\left\{\left[\tfrac{1}{2}(2m+1)\right]^2 B_{11} + (2m+1)v\right\}$$
$$\times \vartheta_k\left(w + \tfrac{1}{2}(2m+1)B_{12}; B_{22}\right), \qquad \text{(A2.10)}$$

$$\vartheta_{3,k} := \sum_{m=-\infty}^{\infty} \exp\{m^2 B_{11} + 2mv\}\vartheta_k(w + mB_{12}; B_{22}),$$

$$\vartheta_{4,k} := \sum_{m=-\infty}^{\infty} (-1)^m \exp\{m^2 B_{11} + 2mv\}\vartheta_k(w + mB_{12}; B_{22}).$$

Selected properties:

$$\vartheta_{3,3}(x_1, x_2; B_{11} \pm i\pi, B_{22} \mp i\pi, B_{12}) = \vartheta_{4,4}(x_1, x_2; B_{11}, B_{22}, B_{12}),$$
$$\vartheta_{2,2}(x_1, x_2; B_{11} \pm i\pi, B_{22} \mp i\pi, B_{12}) = \vartheta_{2,2}(x_1, x_2; B_{11}, B_{22}, B_{12}), \qquad \text{(A2.11)}$$

$$\vartheta_{4,4}\left(x_1 \pm i\frac{\pi}{2}, x_2; B_{11}, B_{22}, B_{12}\right) = \vartheta_{3,4}(x_1, x_2; B_{11}, B_{22}, B_{12}),$$
$$\vartheta_{4,4}\left(x_1, x_2 \pm i\frac{\pi}{2}; B_{11}, B_{22}, B_{12}\right) = \vartheta_{4,3}(x_1, x_2; B_{11}, B_{22}, B_{12}),$$
$$\vartheta_{2,2}\left(x_1 \pm i\frac{\pi}{2}, x_2; B_{11}, B_{22}, B_{12}\right) = \pm i\,\vartheta_{1,2}(x_1, x_2; B_{11}, B_{22}, B_{12}), \qquad \text{(A2.12)}$$
$$\vartheta_{2,2}\left(x_1, x_2 \pm i\frac{\pi}{2}; B_{11}, B_{22}, B_{12}\right) = \pm i\,\vartheta_{2,1}(x_1, x_2; B_{11}, B_{22}, B_{12}).$$

Separation property:

$$\vartheta_{i,k}(v, w; B_{11}, B_{22}, B_{12} = 0) = \vartheta_i(v; B_{11})\vartheta_k(w; B_{22}). \qquad \text{(A2.13)}$$

A2.3 Elliptic integrals and functions, relations to theta functions

Elliptic integrals of the first and second kinds:

$$F(\varphi, k) := \int_0^y \frac{dt}{\sqrt{(1-t^2)(1-k^2t^2)}} = \int_0^\varphi \frac{d\vartheta}{\sqrt{1-k^2 \sin^2 \vartheta}},$$
$$E(\varphi, k) := \int_0^y \sqrt{\frac{1-k^2t^2}{1-t^2}} dt = \int_0^\varphi \sqrt{1-k^2 \sin^2 \vartheta}\, d\vartheta. \quad (A2.14)$$

The number k is called the modulus whereas $k' := \sqrt{1-k^2}$ is referred to as the complementary modulus.

Complete elliptic integrals:

$$K(k) := F\left(\frac{\pi}{2}, k\right), \qquad E(k) := E\left(\frac{\pi}{2}, k\right). \quad (A2.15)$$

Legendre's relation:

$$E(k)K(k') + E(k')K(k) - K(k)K(k') = \frac{\pi}{2}. \quad (A2.16)$$

Jacobian elliptic functions: The elliptic function $\mathrm{sn}(u, k)$ was introduced as the inverse function of the elliptic integral of the first kind. The functions cn and dn are closely related to sn.

$$u = \int_0^y \frac{dt}{\sqrt{(1-t^2)(1-k^2t^2)}} = \int_0^\varphi \frac{d\vartheta}{\sqrt{1-k^2 \sin^2 \theta}} = F(\varphi, k), \quad (A2.17)$$

$$\mathrm{sn}(u, k) := y = \sin \varphi, \qquad \mathrm{am}(u, k) := \varphi,$$
$$\mathrm{cn}(u, k) := \sqrt{1 - y^2} = \cos \varphi, \quad (A2.18)$$
$$\mathrm{dn}(u, k) := \sqrt{1 - k^2 y^2} = \sqrt{1 - k^2 \sin^2 \varphi}.$$

Thus we have the basic identities

$$\begin{aligned}
\mathrm{sn}^2(u, k) + \mathrm{cn}^2(u, k) &= 1, \\
k^2 \mathrm{sn}^2(u, k) + \mathrm{dn}^2(u, k) &= 1, \\
\mathrm{dn}^2(u, k) - k^2 \mathrm{cn}^2(u, k) &= k'^2, \\
k'^2 \mathrm{sn}^2(u, k) + \mathrm{cn}^2(u, k) &= \mathrm{dn}^2(u, k).
\end{aligned} \quad (A2.19)$$

The following relations to the Jacobian theta functions are valid:

$$\operatorname{sn}(-iu, k) = -\frac{i}{\sqrt{k}} \frac{\vartheta_1\left(\frac{\pi u}{2K(k)}; -\pi \frac{K(k')}{K(k)}\right)}{\vartheta_4\left(\frac{\pi u}{2K(k)}; -\pi \frac{K(k')}{K(k)}\right)} = -\frac{i}{\sqrt{k}} \frac{\Theta_1(u, k)}{\Theta_4(u, k)}, \quad (A2.20)$$

$$\operatorname{cn}(-iu, k) = \sqrt{\frac{k'}{k}} \frac{\vartheta_2\left(\frac{\pi u}{2K(k)}; -\pi \frac{K(k')}{K(k)}\right)}{\vartheta_4\left(\frac{\pi u}{2K(k)}; -\pi \frac{K(k')}{K(k)}\right)} = \sqrt{\frac{k'}{k}} \frac{\Theta_2(u, k)}{\Theta_4(u, k)}, \quad (A2.21)$$

$$\operatorname{dn}(-iu, k) = \sqrt{k'} \frac{\vartheta_3\left(\frac{\pi u}{2K(k)}; -\pi \frac{K(k')}{K(k)}\right)}{\vartheta_4\left(\frac{\pi u}{2K(k)}; -\pi \frac{K(k')}{K(k)}\right)} = \sqrt{k'} \frac{\Theta_3(u, k)}{\Theta_4(u, k)}. \quad (A2.22)$$

Heuman's lambda function:

$$\Lambda_0(\psi, k) := \frac{2}{\pi} \left[E(k) F(\psi, k') + K(k) E(\psi, k') - K(k) F(\psi, k') \right]. \quad (A2.23)$$

Jacobian zeta function:

$$Z(u, k) := E(\beta, k) - \frac{E(k)}{K(k)} F(\beta, k) \qquad [\text{with: } \beta \equiv \operatorname{am}(u, k)]. \quad (A2.24)$$

A2.4 Selected derivatives

Derivatives of the elliptic integral of the first kind:

$$\begin{aligned} \frac{\partial}{\partial \varphi} F(\varphi, k) &= \frac{1}{\sqrt{1 - k^2 \sin^2 \varphi}}, \\ \frac{\partial}{\partial k} F(\varphi, k) &= \frac{E(\varphi, k) - k'^2 F(\varphi, k)}{k k'^2} - \frac{k \sin \varphi \cos \varphi}{k'^2 \sqrt{1 - k^2 \sin^2 \varphi}}. \end{aligned} \quad (A2.25)$$

Derivatives of the elliptic integral of the second kind:

$$\begin{aligned} \frac{\partial}{\partial \varphi} E(\varphi, k) &= \sqrt{1 - k^2 \sin^2 \varphi}, \\ \frac{\partial}{\partial k} E(\varphi, k) &= \frac{E(\varphi, k) - F(\varphi, k)}{k}. \end{aligned} \quad (A2.26)$$

In the next equations we use the more concise notation

$$\operatorname{am} u \equiv \operatorname{am}(u, k), \operatorname{sn} u \equiv \operatorname{sn}(u, k), \ldots \quad \text{and} \quad E(u) \equiv E(\operatorname{am}(u, k), k).$$

Derivatives of the Jacobian elliptic functions with respect to the argument:

$$\frac{\partial}{\partial u} \operatorname{am} u = \operatorname{dn} u,$$
$$\frac{\partial}{\partial u} \operatorname{sn} u = \operatorname{cn} u \operatorname{dn} u,$$
$$\frac{\partial}{\partial u} \operatorname{cn} u = -\operatorname{sn} u \operatorname{dn} u,$$
$$\frac{\partial}{\partial u} \operatorname{dn} u = -k^2 \operatorname{sn} u \operatorname{cn} u.$$
(A2.27)

Derivatives of the Jacobian elliptic functions with respect to the modulus:

$$\frac{\partial}{\partial k} \operatorname{am} u = \frac{\operatorname{dn} u}{k\, k'^2} \left[-\mathrm{E}(u) + k'^2 u + k^2 \operatorname{sn} u \frac{\operatorname{cn} u}{\operatorname{dn} u} \right],$$
$$\frac{\partial}{\partial k} \operatorname{sn} u = \frac{\operatorname{dn} u \operatorname{cn} u}{k\, k'^2} \left[-\mathrm{E}(u) + k'^2 u + k^2 \operatorname{sn} u \frac{\operatorname{cn} u}{\operatorname{dn} u} \right],$$
$$\frac{\partial}{\partial k} \operatorname{cn} u = \frac{\operatorname{sn} u \operatorname{dn} u}{k\, k'^2} \left[\mathrm{E}(u) - k'^2 u - k^2 \operatorname{sn} u \frac{\operatorname{cn} u}{\operatorname{dn} u} \right],$$
$$\frac{\partial}{\partial k} \operatorname{dn} u = \frac{k \operatorname{sn} u \operatorname{cn} u}{k'^2} \left[\mathrm{E}(u) - k'^2 u - \operatorname{dn} u \frac{\operatorname{sn} u}{\operatorname{cn} u} \right].$$
(A2.28)

Derivative of $\ln \vartheta_2$ with respect to the argument:

$$\frac{\partial}{\partial w} \ln \left[\vartheta_2 \left(\frac{\pi \mathrm{F}(\beta, k')}{2\mathrm{K}(k)}, -\pi \frac{\mathrm{K}(k')}{\mathrm{K}(k)} \right) \right] = \Lambda_0(\beta, k) \qquad (A2.29)$$

with

$$w = \frac{\pi u}{2\mathrm{K}(k)} = \frac{\pi \mathrm{F}(\beta, k')}{2\mathrm{K}(k)}, \quad \beta = \operatorname{am}(u, k').$$

Appendix 3
Multipole moments of the rotating disc of dust

Here we provide all the quantities that are necessary for calculating the first eleven normalized multipole moments $\tilde{P}_0, \tilde{P}_1, \tilde{P}_2, \ldots, \tilde{P}_{10}$ of the rigidly rotating disc of dust. Their exact definition and the method of how they can be obtained from the axis potential (2.256) were discussed in detail in Subsection 2.3.4. The moments $\tilde{P}_n$ are given as functions of the parameter μ and the more precise structure is

$$\tilde{P}_n(\mu) = \tilde{P}_n\left[b_0(\mu), \Omega_0(\mu), c_j(\mu) : j < j_{\max}(n)\right], \quad (A3.1)$$

where $j_{\max}(n)$ refers to the finite maximal value of j for a given n. The reader could calculate $\tilde{P}_0, \ldots, \tilde{P}_{10}$ using only the definitions to follow. For clarity, we shall make some remarks concerning their derivation and, for the first seven multipole moments, we list the explicit formulae.

A3.1 Definitions and auxiliary coefficients

We rewrite the equations for the parameter functions b_0 (2.202) and Ω_0 (2.204):

$$b_0(\mu) = -\frac{1}{h(\mu)} \operatorname{sn}\left[\hat{I}(\mu), h'(\mu)\right] \operatorname{dn}\left[\hat{I}(\mu), h'(\mu)\right],$$

$$\Omega_0 \equiv \Omega \varrho_0(\mu) = \frac{1}{2}\sqrt{1 - \frac{h'^2(\mu)}{h^2(\mu)}} \operatorname{cn}\left[\hat{I}(\mu), h'(\mu)\right]. \quad (A3.2)$$

The moduli h and h' of the Jacobi elliptic functions sn, cn and dn are given by

$$h(\mu) = \sqrt{\frac{1}{2}\left(1 + \frac{\mu}{\sqrt{1+\mu^2}}\right)}, \quad h'(\mu) = \sqrt{\frac{1}{2}\left(1 - \frac{\mu}{\sqrt{1+\mu^2}}\right)}, \quad (A3.3)$$

and the main argument of these functions is defined as

$$\hat{I}(\mu) := \sqrt[4]{1+\mu^2}\, I_0(\mu), \quad (A3.4)$$

where I_0 is defined as the first of the I_n $(n = 0, 1, 2, \ldots)$

$$I_n(\mu) := \frac{1}{\pi} \int_0^\mu \frac{\ln(\sqrt{1+x^2}+x)}{\sqrt{1+x^2}} \frac{x^n}{\sqrt{\mu-x}} \, dx \,. \tag{A3.5}$$

Furthermore the abbreviations

$$\tau := \sqrt[4]{1 + \frac{1}{\mu^2}} \quad \text{and} \tag{A3.6}$$

$$am := \mathrm{am}\left[\hat{I}(\mu), h'(\mu)\right], \quad sn := \mathrm{sn}\left[\hat{I}(\mu), h'(\mu)\right], \tag{A3.7}$$

$$cn := \mathrm{cn}\left[\hat{I}(\mu), h'(\mu)\right], \quad dn := \mathrm{dn}\left[\hat{I}(\mu), h'(\mu)\right], \tag{A3.8}$$

$$scd := sn \, cn \, dn \,, \quad E := \mathrm{E}(am, h') \tag{A3.9}$$

are used in order to render the equations more concise.

We now provide expressions for the functions c_j up to $j = 11$. These quantities are the coefficients of the function $N(\mu, y)$ (2.228) at infinity, which are needed for calculating the first eleven multipole moments. The first coefficient c_1 is evaluated in Subsection 2.3.4. This is the only one which is necessary for calculating the mass and angular momentum (the first two moments). All this is explained in detail in the aforementioned subsection.

The coefficients c_n of the function N (see (2.228) and (2.257)):

$c_1 = 2 E \tau + (-(\mu (1 + \tau^2) I_0) + I_1)/\sqrt{\mu}$

$c_3 = (-2 E \mu^{3/2} \tau + \mu^{3/2} scd (\tau - \tau^3) + \mu^2 \tau^2 (1 + \tau^2) I_0 - 3 \mu I_1$
$\qquad + 3 I_2)/(3 \mu^{3/2})$

$c_5 = (-2 E \mu^{5/2} \tau (-3 + \tau^4) + \mu^{5/2} scd \, \tau (-1 + \tau^2) (4 + cn^2 (-1 + \tau^2))$
$\qquad + \mu^3 (5 - 3 \tau^2 - 7 \tau^4 + \tau^6) I_0 + 5 \mu^2 (1 + \tau^4) I_1$
$\qquad - 20 \mu I_2 + 10 I_3)/(10 \mu^{5/2})$

$c_7 = (4 E \mu^{7/2} \tau (-5 + 3 \tau^4) - \mu^{7/2} scd \, \tau (-1 + \tau^2) (6 cn^2 (-1 + \tau^2)$
$\qquad + cn^4(-1 + \tau^2)^2 - 3(-5 + \tau^4)) - 2 \mu^4 (14 - 5 \tau^2 - 17 \tau^4 + 3 \tau^6 + \tau^8) I_0$
$\qquad + 14 \mu^3 (1 - 3 \tau^4) I_1 + 14 \mu^2 (5 + \tau^4) I_2 - 84 \mu I_3 + 28 I_4)/(28 \mu^{7/2})$

A3.1 Definitions and auxiliary coefficients

$$c_9 = (2\,E\,\mu^{9/2}\,\tau\,(35 - 30\,\tau^4 + 3\,\tau^8) + \mu^{9/2}\,scd\,\tau\,(-1 + \tau^2)\,(56 - 24\,\tau^4$$
$$+ 8\,cn^4\,(-1 + \tau^2)^2 + cn^6\,(-1 + \tau^2)^3 - 4\,cn^2\,(7 - 7\,\tau^2 - \tau^4 + \tau^6))$$
$$+ \mu^5\,(117 - 35\,\tau^2 - 146\,\tau^4 + 30\,\tau^6 + 21\,\tau^8 - 3\,\tau^{10})\,I_0$$
$$- 9\,\mu^4\,(17 - 26\,\tau^4 + \tau^8)\,I_1 - 144\,\mu^3\,(1 + \tau^4)\,I_2 + 36\,\mu^2\,(11 + \tau^4)\,I_3$$
$$- 288\,\mu\,I_4 + 72\,I_5)/(72\,\mu^{9/2})$$

$$c_{11} = (\mu^{11/2}\,\tau\,(-4\,E\,(63 - 70\,\tau^4 + 15\,\tau^8) - scd\,(-1 + \tau^2)\,(10\,cn^6\,(-1 + \tau^2)^3$$
$$+ cn^8\,(-1 + \tau^2)^4 - 5\,cn^4\,(-1 + \tau^2)^2\,(-9 + \tau^4)$$
$$- 40\,cn^2\,(3 - 3\,\tau^2 - \tau^4 + \tau^6) + 10\,(21 - 14\,\tau^4 + \tau^8)))$$
$$+ 2\,\mu^6\,(-220 + 63\,\tau^2 + 299\,\tau^4 - 70\,\tau^6 - 74\,\tau^8 + 15\,\tau^{10} + 3\,\tau^{12})\,I_0$$
$$+ 22\,\mu^5\,(37 - 50\,\tau^4 + 5\,\tau^8)\,I_1 - 22\,\mu^4\,(1 - 42\,\tau^4 + \tau^8)\,I_2$$
$$- 440\,\mu^3\,(3 + \tau^4)\,I_3 + 88\,\mu^2\,(19 + \tau^4)\,I_4 - 880\,\mu\,I_5 + 176\,I_6)/(176\,\mu^{11/2}).$$

The coefficients $\tilde{m}_n$ of the function g (see (2.254) and (2.255)):

$$\tilde{m}_0 = -b_0 - \Omega_0\,c_1$$

$$\tilde{m}_1 = -i\,(b_0 + 2\,\Omega_0\,c_1)$$

$$\tilde{m}_2 = b_0 + (2 + b_0^2)\,\Omega_0\,c_1 + b_0\,\Omega_0^2\,c_1^2 + (\Omega_0^3\,(c_1^3 - 12\,c_3))/3$$

$$\tilde{m}_3 = (i/3)\,(3\,b_0^3 + 12\,b_0^2\,\Omega_0\,c_1 + 3\,b_0\,\Omega_0^2\,(4 + 3\,c_1^2)$$
$$+ 2\,\Omega_0^3\,(12\,c_1 + c_1^3 - 12\,c_3))$$

$$\tilde{m}_4 = b_0 - 2\,b_0^3 + (2 - 7\,b_0^2)\,\Omega_0\,c_1 - b_0\,\Omega_0^2\,(8 + (5 + b_0^2)\,c_1^2)$$
$$- (2\,\Omega_0^3\,(24\,c_1 + (1 + 2\,b_0^2)\,c_1^3 - 6\,(2 + b_0^2)\,c_3))/3$$
$$+ b_0\,\Omega_0^4\,((-2\,c_1^4)/3 + 8\,c_1\,c_3) + \Omega_0^5\,((-2\,c_1^5)/15 + 4\,c_1^2\,c_3 - 16\,c_5)$$

$$\tilde{m}_5 = (i/15)\,(15\,b_0 - 30\,b_0^3 - 30\,(-1 + 3\,b_0^2 + b_0^4)\,\Omega_0\,c_1$$
$$- 30\,b_0\,\Omega_0^2\,(2\,(1 + c_1^2) + b_0^2\,(2 + 3\,c_1^2))$$
$$- 40\,\Omega_0^3\,((3 + 6\,b_0^2)\,c_1 + 2\,b_0^2\,c_1^3 - 6\,b_0^2\,c_3)$$
$$+ 30\,b_0\,\Omega_0^4\,(-10 + 2\,\tau^4 - 6\,c_1^2 - c_1^4 + 12\,c_1\,c_3)$$
$$- 4\,\Omega_0^5\,(-30\,(-5 + \tau^4)\,c_1 + 10\,c_1^3 + c_1^5 - 30\,c_1^2\,c_3 - 120\,(c_3 - c_5)))$$

$$\tilde{m}_6 = b_0^5 + b_0^2 (-1 + 8 b_0^2) \Omega_0 c_1 + b_0 \Omega_0^2 (-8 - c_1^2 + 16 b_0^2 (1 + c_1^2))$$
$$+ (\Omega_0^3 (24 (-2 + 7 b_0^2) c_1 + (-2 + 34 b_0^2 + 3 b_0^4) c_1^3 + 12 (2 - 7 b_0^2) c_3))/3$$
$$+ (b_0 \Omega_0^4 (168 - 24 \tau^4 + 120 c_1^2 + 5 (2 + b_0^2) c_1^4 - 24 (5 + b_0^2) c_1 c_3))/3$$
$$+ (\Omega_0^5 (-240 (-7 + \tau^4) c_1 + 80 c_1^3 + (4 + 17 b_0^2) c_1^5 - 120 (1 + 2 b_0^2) c_1^2 c_3$$
$$+ 240 (-4 c_3 + (2 + b_0^2) c_5)))/15$$
$$+ b_0 \Omega_0^6 ((17 c_1^6)/45 - (32 c_1^3 c_3)/3 + 16 c_3^2 + 32 c_1 c_5)$$
$$+ \Omega_0^7 ((17 c_1^7)/315 - (8 c_1^4 c_3)/3 + 16 c_1 c_3^2 + 16 c_1^2 c_5 - 64 c_7)$$

$$\tilde{m}_7 = (i/315) (315 b_0 - 945 b_0^3 + 945 b_0^5 + 630 (1 - 5 b_0^2 + 8 b_0^4) \Omega_0 c_1$$
$$+ 315 b_0 \Omega_0^2 (-16 - 7 c_1^2 + 3 b_0^4 c_1^2 + b_0^2 (28 + 25 c_1^2))$$
$$+ 210 \Omega_0^3 (12 (-4 + 11 b_0^2 + b_0^4) c_1 + (-1 + 21 b_0^2 + 13 b_0^4) c_1^3$$
$$- 12 (-1 + 3 b_0^2 + b_0^4) c_3)$$
$$- 105 b_0 \Omega_0^4 (4 (-39 + 3 \tau^4 - 45 c_1^2 - 2 c_1^4 + 24 c_1 c_3)$$
$$+ 3 b_0^2 (-20 + 4 \tau^4 - 24 c_1^2 - 9 c_1^4 + 48 c_1 c_3))$$
$$+ 84 \Omega_0^5 (-30 (-13 + \tau^4 + 2 b_0^2(-5 + \tau^4))c_1 + 10(1 + 8 b_0^2) c_1^3 + 17 b_0^2 c_1^5$$
$$- 240 b_0^2 c_1^2 c_3 - 120 (c_3 + 2 b_0^2 c_3 - 2 b_0^2 c_5))$$
$$+ 21 b_0 \Omega_0^6 (900 c_1^2 + 120 c_1^4 + 17 c_1^6 - 180 \tau^4 (4 + c_1^2) - 480 c_1^3 c_3$$
$$+ 240 (7 + 3 c_3^2) - 1440 c_1 (c_3 - c_5))$$
$$+ 2 \Omega_0^7 (-420 (-5 + \tau^4) c_1^3 + 168 c_1^5 + 17 c_1^7$$
$$- 840 c_1^4 c_3 - 5040 c_1 (-7 + 3 \tau^4 - c_3^2)$$
$$- 5040 c_1^2 (c_3 - c_5) + 5040 ((-5 + \tau^4) c_3 + 4 (c_5 - c_7))))$$

$$\tilde{m}_8 = -b_0 + 4 b_0^3 - 4 b_0^5 - (2 - 12 b_0^2 + 16 b_0^4 + 3 b_0^6) \Omega_0 c_1$$
$$- b_0 \Omega_0^2 (-8 (1 + c_1^2) + b_0^4 (8 + 17 c_1^2) + b_0^2 (16 + 19 c_1^2))$$
$$- (4 \Omega_0^3(6 (-2 + 5 b_0^2 + 8 b_0^4) c_1 + b_0^2(5 + 23 b_0^2)c_1^3 + 3 b_0^2(1 - 8 b_0^2)c_3))/3$$
$$- (b_0 \Omega_0^4 (-120 + 24 \tau^4 + 72 c_1^2 - 2 c_1^4 + 3 b_0^4 c_1^4 + 24 c_1 c_3$$
$$+ b_0^2 (336 - 48 \tau^4 + 384 c_1^2 + 74 c_1^4 - 384 c_1 c_3)))/3$$
$$- (\Omega_0^5 (-120 (7 b_0^2 (-7 + \tau^4) - 2 (-5 + \tau^4)) c_1 + 80 (-1 + 17 b_0^2) c_1^3$$
$$+ (-4 + 149 b_0^2 + 30 b_0^4) c_1^5 - 60 (-2 + 34 b_0^2 + 3 b_0^4) c_1^2 c_3$$

$$- 240\,(-2 + 7\,b_0^2)\,(2\,c_3 - c_5)))/15$$
$$- (b_0\,\Omega_0^6\,(12600\,c_1^2 + 1200\,c_1^4 + (85 + 77\,b_0^2)\,c_1^6 - 360\,\tau^4\,(16 + 5\,c_1^2)$$
$$- 1200\,(2 + b_0^2)\,c_1^3\,c_3 + 720\,(24 + (5 + b_0^2)\,c_3^2)$$
$$- 1440\,c_1\,(10\,c_3 - (5 + b_0^2)\,c_5)))/45$$
$$- (2\,\Omega_0^7\,(-840\,(-7 + \tau^4)\,c_1^3 + 336\,c_1^5 + (17 + 124\,b_0^2)\,c_1^7$$
$$- 210\,(4 + 17\,b_0^2)\,c_1^4\,c_3$$
$$- 5040\,c_1\,(-24 + 8\,\tau^4 - (1 + 2\,b_0^2)\,c_3^2) - 5040\,c_1^2\,(2\,c_3 - (1 + 2\,b_0^2)\,c_5)$$
$$- 10080\,(-((-7 + \tau^4)\,c_3) - 4\,c_5 + (2 + b_0^2)\,c_7)))/315$$
$$+ b_0\,\Omega_0^8\,((-62\,c_1^8)/315 + (136\,c_1^5\,c_3)/15 - 64\,c_1^2\,c_3^2 - (128\,c_1^3\,c_5)/3$$
$$+ 128\,c_3\,c_5 + 128\,c_1\,c_7)$$
$$+ \Omega_0^9\,((-62\,c_1^9)/2835 + (68\,c_1^6\,c_3)/45 - (64\,c_1^3\,c_3^2)/3$$
$$- (32\,c_1^4\,c_5)/3 + 128\,c_1\,c_3\,c_5 + 64\,c_1^2\,c_7 + (64\,(c_3^3 - 12\,c_9))/3)$$

$$\tilde{m}_9 = (i/2835)\,(2835\,b_0^3 - 2835\,b_0^5 - 2835\,b_0^7 - 5670\,b_0^2\,(-1 - b_0^2 + 7\,b_0^4)\,\Omega_0\,c_1$$
$$- 2835\,b_0\,\Omega_0^2\,(12 - c_1^2 + 12\,b_0^4\,(3 + 4\,c_1^2) - b_0^2\,(32 + 11\,c_1^2))$$
$$- 1890\,\Omega_0^3\,(12\,(3 - 14\,b_0^2 + 25\,b_0^4)\,c_1 + (1 - 17\,b_0^2 + 98\,b_0^4 + 6\,b_0^6)\,c_1^3$$
$$- 12\,(1 - 5\,b_0^2 + 8\,b_0^4)\,c_3)$$
$$- 945\,b_0\,\Omega_0^4\,(36\,b_0^4\,c_1\,(c_1 + c_1^3 - 2\,c_3)$$
$$+ b_0^2\,(804 - 84\,\tau^4 + 972\,c_1^2 + 125\,c_1^4 - 600\,c_1\,c_3)$$
$$+ 2\,(-240 + 24\,\tau^4 - 120\,c_1^2 - 7\,c_1^4 + 84\,c_1\,c_3))$$
$$- 378\,\Omega_0^5\,(-60\,(40 - 4\,\tau^4 + b_0^4\,(-5 + \tau^4) + b_0^2\,(-107 + 11\,\tau^4))\,c_1$$
$$+ 20\,(-4 + 71\,b_0^2 + 13\,b_0^4)\,c_1^3 + (-2 + 96\,b_0^2 + 107\,b_0^4)\,c_1^5$$
$$- 60\,(-1 + 21\,b_0^2 + 13\,b_0^4)\,c_1^2\,c_3$$
$$- 240\,((-4 + 11\,b_0^2 + b_0^4)\,c_3 - (-1 + 3\,b_0^2 + b_0^4)\,c_5))$$
$$+ 126\,b_0\,\Omega_0^6(2\,(-6615\,c_1^2 - 450\,c_1^4 - 17\,c_1^6 + 45\,\tau^4\,(28 + 15\,c_1^2) + 480\,c_1^3\,c_3$$
$$- 180\,(31 + 4\,c_3^2) + 360\,c_1\,(15\,c_3 - 4\,c_5))$$
$$+ 3\,b_0^2\,(-900\,c_1^2 - 270\,c_1^4 - 67\,c_1^6 + 180\,\tau^4\,(2 + c_1^2) + 1080\,c_1^3\,c_3$$
$$- 120\,(7 + 6\,c_3^2) + 1440\,c_1\,(c_3 - c_5)))$$

$$-72\,\Omega_0^7\,(-105\,(-13+\tau^4+8\,b_0^2\,(-5+\tau^4))\,c_1^3+42\,(1+17\,b_0^2)\,c_1^5$$
$$+124\,b_0^2\,c_1^7-3570\,b_0^2\,c_1^4\,c_3$$
$$-1260\,c_1\,(-31+7\,\tau^4+2\,b_0^2\,(-7+3\,\tau^4-4\,c_3^2))$$
$$-1260\,c_1^2\,(c_3+8\,b_0^2\,c_3-8\,b_0^2\,c_5)$$
$$+1260\,((-13+\tau^4+2\,b_0^2\,(-5+\tau^4))\,c_3+4\,(c_5+2\,b_0^2\,c_5-2\,b_0^2\,c_7)))$$
$$-54\,b_0\,\Omega_0^8\,(1680\,\tau^8+2100\,c_1^4+238\,c_1^6+31\,c_1^8-1428\,c_1^5\,c_3$$
$$-420\,\tau^4\,(56+18\,c_1^2+c_1^4-12\,c_1\,c_3)+2520\,c_1^2\,(7+4\,c_3^2)$$
$$-6720\,c_1^3\,(c_3-c_5)+5040\,(7+2\,c_3^2-4\,c_3\,c_5)$$
$$-5040\,c_1\,(5\,c_3-4\,c_5+4\,c_7))$$
$$-4\,\Omega_0^9\,(-756\,(-5+\tau^4)\,c_1^5+306\,c_1^7+31\,c_1^9-2142\,c_1^6\,c_3$$
$$-7560\,c_1^3\,(-7+3\,\tau^4-4\,c_3^2)-15120\,c_1^4\,(c_3-c_5)$$
$$+45360\,c_1\,(21-14\,\tau^4+\tau^8+2\,c_3^2-4\,c_3\,c_5)$$
$$+22680\,c_1^2\,((-5+\tau^4)\,c_3+4\,(c_5-c_7))+30240\,(3\,(-7+3\,\tau^4)\,c_3-c_3^3$$
$$-3\,((-5+\tau^4)\,c_5+4\,(c_7-c_9)))))$$

$$\tilde{m}_{10}=-b_0+4\,b_0^3-6\,b_0^5+4\,b_0^7+(-2+13\,b_0^2-32\,b_0^4+32\,b_0^6)\,\Omega_0\,c_1$$
$$+b_0\,\Omega_0^2\,(24+9\,c_1^2+6\,b_0^6\,c_1^2-10\,b_0^2\,(8+5\,c_1^2)+8\,b_0^4\,(9+10\,c_1^2))$$
$$+(\Omega_0^3\,(24\,(6-31\,b_0^2+40\,b_0^4+3\,b_0^6)\,c_1+(2-84\,b_0^2+253\,b_0^4+90\,b_0^6)\,c_1^3$$
$$-12\,(2-12\,b_0^2+16\,b_0^4+3\,b_0^6)\,c_3))/3$$
$$-(b_0\,\Omega_0^4\,(-8\,(-57+3\,\tau^4-63\,c_1^2-2\,c_1^4+24\,c_1\,c_3)$$
$$+b_0^4\,(-168+24\,\tau^4-408\,c_1^2-163\,c_1^4+408\,c_1\,c_3)$$
$$+b_0^2\,(-816+48\,\tau^4-1296\,c_1^2-119\,c_1^4+456\,c_1\,c_3)))/3$$
$$+(\Omega_0^5\,(-120\,(38-2\,\tau^4+8\,b_0^4\,(-7+\tau^4)+b_0^2\,(-97+5\,\tau^4))\,c_1$$
$$+80\,(-1+37\,b_0^2+46\,b_0^4)\,c_1^3+b_0^2\,(103+736\,b_0^2+15\,b_0^4)\,c_1^5$$
$$-240\,b_0^2\,(5+23\,b_0^2)\,c_1^2\,c_3-240\,(2\,(-2+5\,b_0^2+8\,b_0^4)\,c_3$$
$$+b_0^2\,(1-8\,b_0^2)\,c_5)))/15$$
$$+(b_0\,\Omega_0^6\,(-4320+22680\,c_1^2+720\,c_1^4-17\,c_1^6-1080\,\tau^4\,(-4+c_1^2)$$
$$-8640\,c_1\,c_3+480\,c_1^3\,c_3-720\,c_3^2+15\,b_0^4\,(7\,c_1^6-48\,c_1^3\,c_3)$$

$$
\begin{aligned}
&- 8\,b_0^2\,(-5040\,c_1^2 - 1110\,c_1^4 - 139\,c_1^6 + 720\,\tau^4\,(2 + c_1^2) + 2220\,c_1^3\,c_3 \\
&- 1440\,(3 + c_3^2) + 2880\,c_1\,(2\,c_3 - c_5)) - 1440\,c_1\,c_5))/45 \\
&+ (2\,\Omega_0^7\,(-840\,(5 - \tau^4 + 17\,b_0^2\,(-7 + \tau^4))\,c_1^3 + 84\,(-4 + 149\,b_0^2)\,c_1^5 \\
&+ (-17 + 1099\,b_0^2 + 378\,b_0^4)\,c_1^7 - 210\,(-4 + 149\,b_0^2 + 30\,b_0^4)\,c_1^4\,c_3 \\
&+ 2520\,c_1\,(3\,b_0^4\,c_3^2 + 2\,(-6 + 6\,\tau^4 - c_3^2) + b_0^2\,(168 - 56\,\tau^4 + 34\,c_3^2)) \\
&- 2520\,c_1^2\,((-4 + 68\,b_0^2)\,c_3 + (2 - 34\,b_0^2 - 3\,b_0^4)\,c_5) \\
&+ 5040\,((7\,b_0^2\,(-7 + \tau^4) - 2\,(-5 + \tau^4))\,c_3 \\
&+ 2\,(-2 + 7\,b_0^2)\,(2\,c_5 - c_7))))/315 \\
&+ (2\,b_0\,\Omega_0^8\,(12600\,\tau^8 + 29400\,c_1^4 + 2380\,c_1^6 + 5\,(31 + 44\,b_0^2)\,c_1^8 \\
&- 84\,(85 + 77\,b_0^2)\,c_1^5\,c_3 - 4200\,\tau^4\,(54 + 24\,c_1^2 + c_1^4 - 12\,c_1\,c_3) \\
&+ 25200\,c_1^2\,(12 + (2 + b_0^2)\,c_3^2) - 16800\,c_1^3\,(4\,c_3 - (2 + b_0^2)\,c_5) \\
&+ 2520\,(165 + 40\,c_3^2 - 8\,(5 + b_0^2)\,c_3\,c_5) \\
&- 10080\,c_1\,(35\,c_3 + 2\,(-10\,c_5 + (5 + b_0^2)\,c_7))))/315 \\
&+ (2\,\Omega_0^9\,(-3024\,(-7 + \tau^4)\,c_1^5 + 1224\,c_1^7 + (62 + 691\,b_0^2)\,c_1^9 \\
&- 252\,(17 + 124\,b_0^2)\,c_1^6\,c_3 - 15120\,c_1^3\,(-24 + 8\,\tau^4 - (4 + 17\,b_0^2)\,c_3^2) \\
&- 7560\,c_1^4\,(8\,c_3 - (4 + 17\,b_0^2)\,c_5) \\
&+ 45360\,c_1\,(165 - 90\,\tau^4 + 5\,\tau^8 + 8\,c_3^2 - 8\,(1 + 2\,b_0^2)\,c_3\,c_5) \\
&+ 90720\,c_1^2\,((-7 + \tau^4)\,c_3 - 2\,(-2\,c_5 + c_7 + 2\,b_0^2\,c_7)) \\
&- 60480\,(-24\,(-3 + \tau^4)\,c_3 + (1 + 2\,b_0^2)\,c_3^3 \\
&- 6\,(-((-7 + \tau^4)\,c_5) - 4\,c_7 + (2 + b_0^2)\,c_9))))/2835 \\
&+ (2\,b_0\,\Omega_0^{10}\,(691\,c_1^{10} - 44640\,c_1^7\,c_3 + 642600\,c_1^4\,c_3^2 + 257040\,c_1^5\,c_5 \\
&- 3628800\,c_1^2\,c_3\,c_5 - 1209600\,c_1^3\,c_7 + 1814400\,(c_5^2 + 2\,c_3\,c_7) \\
&- 1209600\,c_1\,(c_3^3 - 3\,c_9)))/14175 \\
&+ \Omega_0^{11}\,((1382\,c_1^{11})/155925 - (248\,c_1^8\,c_3)/315 + (272\,c_1^5\,c_3^2)/15 \\
&+ (272\,c_1^6\,c_5)/45 - (512\,c_1^3\,c_3\,c_5)/3 - (128\,c_1^4\,c_7)/3 \\
&+ 256\,c_1\,(c_5^2 + 2\,c_3\,c_7) - (256\,c_1^2\,(c_3^3 - 3\,c_9))/3 \\
&+ 256\,(c_3^2\,c_5 - 4\,c_{11})).
\end{aligned}
$$

A3.2 Normalized multipole moments

With the normalized coefficients $\tilde{m}_n$, we are now able to calculate the first six normalized mass moments $\tilde{M}_0, \tilde{M}_2, \tilde{M}_4, \tilde{M}_6, \tilde{M}_8$ and $\tilde{M}_{10}$ and the first five normalized rotational moments $\tilde{J}_1, \tilde{J}_3, \tilde{J}_5, \tilde{J}_7$ and $\tilde{J}_9$. The scheme derived in Fodor *et al.* (1989) and adapted to our notation and normalization reads as follows.

Scheme for $\tilde{M}_n(\tilde{m}_i)$ and $\tilde{J}_n(\tilde{m}_i)$ up to $n = 10$:

$$\tilde{M}_0 = \tilde{m}_0,$$

$$\tilde{J}_1 = \Im\,\tilde{m}_1,$$

$$\tilde{M}_2 = -\tilde{m}_2,$$

$$\tilde{J}_3 = -\Im\,\tilde{m}_3,$$

$$\tilde{M}_4 = \tilde{m}_4 - \frac{1}{7}(\tilde{m}_2\tilde{m}_0 - \tilde{m}_1^2)\,\overline{\tilde{m}_0},$$

$$\tilde{J}_5 = \Im\left[\tilde{m}_5 - \frac{1}{21}(\tilde{m}_2\tilde{m}_0 - \tilde{m}_1^2)\,\overline{\tilde{m}_1} - \frac{1}{3}(\tilde{m}_3\tilde{m}_0 - \tilde{m}_2\tilde{m}_1)\,\overline{\tilde{m}_0}\right],$$

$$\tilde{M}_6 = -\left[\tilde{m}_6 + \frac{1}{33}(\tilde{m}_2\tilde{m}_0 - \tilde{m}_1^2)\,\overline{\tilde{m}_0}^2\,\tilde{m}_0 - \frac{5}{231}(\tilde{m}_2\tilde{m}_0 - \tilde{m}_1^2)\,\overline{\tilde{m}_2}\right.$$

$$\left. - \frac{4}{33}(\tilde{m}_3\tilde{m}_0 - \tilde{m}_2\tilde{m}_1)\,\overline{\tilde{m}_1} - \frac{8}{33}(\tilde{m}_3\tilde{m}_1 - \tilde{m}_2^2)\,\overline{\tilde{m}_0}\right.$$

$$\left. - \frac{6}{11}(\tilde{m}_4\tilde{m}_0 - \tilde{m}_3\tilde{m}_1)\,\overline{\tilde{m}_0}\right],$$

$$\tilde{J}_7 = -\Im\left[\tilde{m}_7 - \ldots\right],$$

$$\tilde{M}_8 = \tilde{m}_8 - \ldots,$$

$$\tilde{J}_9 = \Im\left[\tilde{m}_9 - \ldots\right],$$

$$\tilde{M}_{10} = -\left[\tilde{m}_{10} - \ldots\right].$$

The complete formulae for $\tilde{J}_7, \tilde{J}_9$ and $\tilde{M}_8, \tilde{M}_{10}$ can be found in Fodor *et al.* (1989). Next we list the resulting expressions for $\tilde{M}_0, \tilde{M}_2, \tilde{M}_4, \tilde{M}_6$ and $\tilde{J}_1, \tilde{J}_3, \tilde{J}_5$.

The first seven multipole moments:

$$\tilde{M}_0 = -b_0 - \Omega_0\,c_1$$

$$\tilde{J}_1 = -b_0 - 2\,\Omega_0\,c_1$$

$$\tilde{M}_2 = (-3\,b_0 - 6\,\Omega_0\,c_1 - 3\,b_0^2\,\Omega_0\,c_1 - 3\,b_0\,\Omega_0^2\,c_1^2 - \Omega_0^3\,c_1^3 + 12\,\Omega_0^3\,c_3)/3$$

A3.2 Normalized multipole moments

$$\tilde{J}_3 = -b_0^3 - 4 b_0^2 \Omega_0 c_1 - b_0 \Omega_0^2 (4 + 3 c_1^2) - (2 \Omega_0^3 (12 c_1 + c_1^3 - 12 c_3))/3$$

$$\begin{aligned}\tilde{M}_4 =\ & b_0 - 2 b_0^3 - ((-14 + 48 b_0^2 + b_0^4) \Omega_0 c_1)/7 \\ & - (2 b_0 \Omega_0^2 (28 + (16 + 5 b_0^2) c_1^2))/7 \\ & - (2 \Omega_0^3 (168 c_1 + (4 + 19 b_0^2) c_1^3 - 12 (7 + 4 b_0^2) c_3))/21 \\ & - (b_0 \Omega_0^4 (19 c_1^4 - 192 c_1 c_3))/21 - (\Omega_0^5 (19 c_1^5 \\ & - 480 c_1^2 c_3 + 1680 c_5))/105\end{aligned}$$

$$\begin{aligned}\tilde{J}_5 =\ & (63 b_0 - 105 b_0^3 - 21 b_0^5 - 6 (-21 + 46 b_0^2 + 38 b_0^4) \Omega_0 c_1 \\ & - 6 b_0 \Omega_0^2 (42 + 16 c_1^2 + b_0^2 (56 + 89 c_1^2)) \\ & - 24 \Omega_0^3 (7 (3 + 8 b_0^2) c_1 + 3 (-1 + 6 b_0^2) c_1^3 - 45 b_0^2 c_3) \\ & + b_0 \Omega_0^4 (-1260 + 252 \tau^4 - 1176 c_1^2 - 145 c_1^4 + 1560 c_1 c_3) \\ & + 63 \Omega_0^5 (8 (-5 + \tau^4) c_1 - (16 c_1^3)/3 - (74 c_1^5)/315 + (160 c_1^2 c_3)/21 \\ & + 32 (c_3 - c_5)))/63\end{aligned}$$

$$\begin{aligned}\tilde{M}_6 =\ & (2310 b_0^3 - 5775 b_0^5 - 15 b_0^2 (-852 + 2462 b_0^2 + 7 b_0^4) \Omega_0 c_1 \\ & - 15 b_0 \Omega_0^2 (110 b_0^4 c_1^2 - 3 (616 + 331 c_1^2) + b_0^2 (4312 + 4348 c_1^2)) \\ & - 5 \Omega_0^3 (336 (-33 + 134 b_0^2) c_1 + (-1278 + 8092 b_0^2 + 1711 b_0^4) c_1^3 \\ & - 12 (-462 + 1594 b_0^2 + 133 b_0^4) c_3) \\ & - 5 b_0 \Omega_0^4 (38808 - 5544 \tau^4 + 33432 c_1^2 + 5 (331 + 494 b_0^2) c_1^4 \\ & - 48 (503 + 223 b_0^2) c_1 c_3) \\ & + 3 \Omega_0^5 (18480 (-7 + \tau^4) c_1 - 8400 c_1^3 - 7 (-26 + 389 b_0^2) c_1^5 \\ & + 120 (35 + 253 b_0^2) c_1^2 c_3 - 1680 (-44 c_3 + (22 + 17 b_0^2) c_5)) \\ & - b_0 \Omega_0^6 (2723 c_1^6 - 60720 c_1^3 c_3 + 43200 c_3^2 + 171360 c_1 c_5) \\ & - \Omega_0^7 (389 c_1^7 - 15180 c_1^4 c_3 + 43200 c_1 c_3^2 + 85680 c_1^2 c_5 \\ & - 221760 c_7))/3465.\end{aligned}$$

The coefficients $\tilde{m}_0, \tilde{m}_1, \ldots, \tilde{m}_{10}$ are sufficient for calculating $\tilde{M}_8, \tilde{M}_{10}$ and $\tilde{J}_7$, $\tilde{J}_9$. Because of the length of the corresponding expressions, we shall not list them here. They were used, of course, for producing the plots in Fig. 2.13. To derive the moments, we made use of a *Mathematica* program.

A3.3 Multipole moments in the extreme relativistic limit

Let us conclude with a remark regarding the extreme relativistic limit $\mu \to \mu_0$. We know from Subsection 2.3.5 that the 'exterior' metric becomes that of the extreme Kerr solution, and therefore the multipole moments tend to

$$\tilde{M}_{2l}(\mu_0) = (-1)^l \, \tilde{m}_{2l}(\mu_0) = 1,$$
$$\tilde{J}_{2l+1}(\mu_0) = (-1)^l \, \Im \, \tilde{m}_{2l+1}(\mu_0) = 1. \tag{A3.10}$$

Based on the expressions given in this appendix, we can see this explicitly for the first eleven moments. In the limit, we have $\Omega_0 \to 0$ and $b_0 \to -1$. For the coefficients $\tilde{m}_0, \tilde{m}_1, \ldots, \tilde{m}_{10}$, we immediately see that

$$\tilde{m}_{2l}(\mu_0) = (-1)^l,$$
$$\tilde{m}_{2l+1}(\mu_0) = (-1)^l \, \mathrm{i}, \tag{A3.11}$$

which can in fact be shown for all l. The structure for the first eleven multipole moments is

$$\tilde{M}_{2l} = (-1)^l \tilde{m}_{2l} + \sum_{s(l)} A_{s(l)} \left[\tilde{m}_{i_{s(l)}} \tilde{m}_{j_{s(l)}} - \tilde{m}_{i_{s(l)}-1} \tilde{m}_{j_{s(l)}+1} \right],$$
$$\tilde{J}_{2l+1} = (-1)^l \Im \tilde{m}_{2l+1} + \sum_{t(l)} B_{s(l)} \left[\tilde{m}_{i_{t(l)}} \tilde{m}_{j_{t(l)}} - \tilde{m}_{i_{t(l)}-1} \tilde{m}_{j_{t(l)}+1} \right]. \tag{A3.12}$$

Using (A3.11), one can verify

$$\left[\tilde{m}_i \tilde{m}_j - \tilde{m}_{i-1} \tilde{m}_{j+1} \right] (\mu_0) = 0, \tag{A3.13}$$

from which it follows that (A3.10) holds.

Appendix 4
The disc solution as a Bäcklund limit

In this appendix, we discuss an alternative representation of the solution for the rigidly rotating disc of dust. The underlying mathematical structure of this formulation is given through the so-called Bäcklund transformation, which is a technique that enables one to construct explicit solutions to the linear matrix problem (2.41) and the corresponding Ernst potentials f. These solutions take a particularly simple form, since they can be written as quotients of determinants in which only elementary functions and functions that can be calculated from a 'seed solution' f_0 appear (see below for examples). The Kerr solution for a rotating black hole in vacuum, Equation (2.358), can be considered as a particular example of a Bäcklund transform, see e.g. Neugebauer (1980a). Moreover, the method allows for the construction of regular Ernst potentials, which correspond to disc-like sources of the gravitational field. In particular, it is possible to identify the rigidly rotating disc of dust as a well-defined limit of these solutions.

After the introduction of disc-like solutions, generated by Bäcklund transformations, depending on a set of parameters as well as a real analytic function, an appropriate generalization is given which allows the Ernst potentials to be written in terms of *two* free functions. For the rigidly rotating disc of dust, these functions take on a simple explicit form.

A4.1 Disc-like solutions of the Bäcklund type

The expression

$$f = f(\varrho/\varrho_0, \zeta/\varrho_0; \{X_\nu\}_q; g) = f_0 \frac{\mathcal{D}_-}{\mathcal{D}_+} \tag{A4.1}$$

satisfies the Ernst equation, where $\{X_1, \ldots, X_q\} =: \{X_\nu\}_q$ is a set of complex parameters, g a real analytic function defined on the interval [0, 1], and $\mathcal{D}_\pm$ and f_0

are given by

$$\mathcal{D}_\pm = \begin{vmatrix} 1 & 1 & 1 & 1 & 1 & \cdots & 1 & 1 \\ \pm 1 & \alpha_1\lambda_1 & \alpha_1^*\lambda_1^* & \alpha_2\lambda_2 & \alpha_2^*\lambda_2^* & \cdots & \alpha_q\lambda_q & \alpha_q^*\lambda_q^* \\ 1 & \lambda_1^2 & (\lambda_1^*)^2 & \lambda_2^2 & (\lambda_2^*)^2 & \cdots & \lambda_q^2 & (\lambda_q^*)^2 \\ \pm 1 & \alpha_1\lambda_1^3 & \alpha_1^*(\lambda_1^*)^3 & \alpha_2\lambda_2^3 & \alpha_2^*(\lambda_2^*)^3 & \cdots & \alpha_q\lambda_q^3 & \alpha_q^*(\lambda_q^*)^3 \\ \vdots & \vdots & \vdots & \vdots & \vdots & \ddots & \vdots & \vdots \\ 1 & \lambda_1^{2q} & (\lambda_1^*)^{2q} & \lambda_2^{2q} & (\lambda_2^*)^{2q} & \cdots & \lambda_q^{2q} & (\lambda_q^*)^{2q} \end{vmatrix} \tag{A4.2}$$

and

$$f_0 = \exp\left(-\int_{-1}^{1} \frac{(-1)^q g(x^2) \, dx}{W_1(ix)}\right) \tag{A4.3}$$

with

$$W_1(X) = \sqrt{(X - \zeta/\varrho_0)^2 + (\varrho/\varrho_0)^2} \quad (\Re(W_1) < 0),$$

$$\lambda_\nu = \sqrt{\frac{\varrho_0 X_\nu - i\bar{z}}{\varrho_0 X_\nu + iz}}, \quad \lambda_\nu^* \bar{\lambda}_\nu = 1 \quad (z = \varrho + i\zeta),$$

$$\alpha_\nu = -\tanh\left(\frac{\lambda_\nu}{2\varrho_0}(\varrho_0 X_\nu + iz)\int_{-1}^{1} \frac{(-1)^q g(x^2)\,dx}{(ix - X_\nu)W_1(ix)}\right), \quad \alpha_\nu^* \bar{\alpha}_\nu = 1.$$

Through the additional requirement that for each parameter X_ν there must also be a parameter X_μ with $X_\nu = -\overline{X}_\mu$, reflectional symmetry, $f(\varrho, -\zeta) = \overline{f(\varrho, \zeta)}$, is ensured.[1] Moreover, the parameters X_ν are assumed to lie outside the imaginary interval $[-i, i]$. The above Ernst potential f is obtained by a multiple Bäcklund transformation applied to the real seed solution f_0, see Neugebauer (1980a).

The particular ansatz chosen for the seed solution f_0 guarantees a resulting Ernst potential that corresponds to a disc-like source of the gravitational field. Furthermore, f does not possess singularities at $(\varrho, \zeta) = \varrho_0(|\Im[X_\nu]|, -\Re[X_\nu])$. This is due to the fact that $\alpha_\nu \lambda_\nu$ is a function of λ_ν^2, and this means that f does not behave like a square root function near the critical points $(\varrho, \zeta) = \varrho_0(|\Im[X_\nu]|, -\Re[X_\nu])$, but rather like a rational function. In addition, one has to make sure that no zeros in the denominator of (A4.1) occur. The real function g that enters the Ernst potential

---

[1] Hence, the set $\{iX_\nu\}_q$ consists of real parameters and/or pairs of complex conjugate parameters.

is assumed to be analytic on [0, 1] in order to guarantee analytic behaviour of the surface energy-momentum distribution. The additional requirement

$$g(1) = 0 \tag{A4.4}$$

ensures regularity at the rim of the disc.

A4.2 Generalization of the Bäcklund type solutions by a limiting process

The set $\{X_\nu\}_q$ of complex parameters can be translated into an analytic function

$$\xi : [0, 1] \to \mathbb{R}$$

such that the corresponding Ernst potential depends on *two* real analytic functions defined on [0, 1]:

$$f = f(\varrho/\varrho_0, \zeta/\varrho_0; \xi; g).$$

This concept proves to be sufficiently general to describe arbitrarily rotating discs. In this manner it becomes possible to describe the solution of the rigidly rotating disc of dust as a well-defined limit of the Bäcklund type solutions.

The following equalities for the above solutions $f = f(\{X_\nu\}_q; g)$ will help in introducing the aforementioned, analytic function ξ, see Ansorg (2001):

$$f[\{X_1, \ldots, X_{q-2}, X_{q-1}, X_q\}; g] = f[\{X_1, \ldots, X_{q-2}\}; g]$$
$$\text{if } X_{q-1} = -X_q \in \mathbb{R}$$

$$f[\{X_1, \ldots, X_{q-2}, X_{q-1}, X_q\}; g] = f[\{X_1, \ldots, X_{q-2}\}; g]$$
$$\text{if } X_{q-1} = \overline{X}_q$$

$$\lim_{t \to \infty} f[\{X_1, \ldots, X_{q-1}, it\}; g] = f[\{X_1, \ldots, X_{q-1}\}; g]$$
$$\text{if } t \in \mathbb{R}$$

$$\lim_{X_q \to \infty} f[\{X_1, \ldots, X_{q-2}, X_{q-1}, X_q\}; g] = f[\{X_1, \ldots, X_{q-2}\}; g]$$
$$\text{if } X_{q-1} = -\overline{X}_q.$$

The desired function $\xi = \xi(\{X_\nu\}_q)$ is supposed to be invariant under the above modifications of the set $\{X_\nu\}_q$ that do not affect the Ernst potential. This requirement is met by the real analytic function

$$\xi(x^2; \{X_\nu\}_q) = \frac{1}{x} \ln \left[\prod_{\nu=1}^{q} \frac{i X_\nu - x}{i X_\nu + x} \right], \quad x \in [-1, 1], \tag{A4.5}$$

which can be proved by considering that for each parameter X_ν there is also a parameter X_μ with $X_\nu = -\overline{X}_\mu$, and that, moreover, the parameters X_ν do not lie on the imaginary interval $[-i, i]$.

The set $\mathcal{X}$ of all functions $\xi = \xi(x^2; \{X_\nu\}_q), q \in \mathbb{N}$, which are defined by (A4.5) forms a dense subset of the set $\mathcal{A}$ of all real analytic functions on $[0, 1]$. Now, for a given function g, each $\xi \in \mathcal{X}$ is mapped by (A4.1) onto a uniquely defined Ernst potential $f \in \mathcal{E}$:

$$\Phi_g : \mathcal{X} \longrightarrow \mathcal{E}, \quad \Phi_g(\xi) = f(\{X_\nu\}_q; g), \tag{A4.6}$$

where the set $\{X_\nu\}_q$ results from ξ by (A4.5).[2]

The mapping Φ_g can be extended to form a continuous function defined on $\mathcal{A}$.[3] It then follows that, given the two real functions g and ξ, defined and analytic on the interval $[0, 1]$, the Ernst potential

$$f(\xi; g) = \lim_{q \to \infty} f(\{X_\nu^{(q)}\}_q; g)$$

exists and is independent of the particular choice of the sequence $\{\{X_\nu^{(q)}\}_q\}_{q=q_0}^\infty$ which serves to represent ξ by

$$\xi(x^2) = \frac{1}{x} \lim_{q \to \infty} \ln \left[\prod_{\nu=1}^{q} \frac{i X_\nu^{(q)} - x}{i X_\nu^{(q)} + x} \right] \quad \text{for} \quad x \in [-1, 1].$$

It can be shown that, in this formulation, the solution for the rigidly rotating disc of dust assumes the form $f = f(\xi; g)$ with the functions ξ and g given by

$$\xi(x^2) = \frac{1}{2x} \ln \frac{x^2 - C_1(\mu)x + C_2(\mu)}{x^2 + C_1(\mu)x + C_2(\mu)},$$

$$C_1(\mu) = \sqrt{2[1 + C_2(\mu)]}, \quad C_2(\mu) = \frac{1}{\mu}\sqrt{1 + \mu^2},$$

$$g(x^2) = -\frac{1}{\pi} \operatorname{arcsinh}[\mu(1 - x^2)].$$

They depend parametrically on μ, $0 < \mu < \mu_0 = 4.62966184\ldots$, which was introduced in (2.79).

Note that a rather technical detail is the determination of an appropriate set $\{X_\nu\}_q$ to give a satisfactory approximation of ξ in terms of (A4.5). There are many ways to do this and we here provide a single, concrete example.

---

[2] Here, $\mathcal{E}$ denotes the set of all Ernst potentials corresponding to disc-like sources.
[3] Detailed mathematical aspects are discussed in Ansorg (2001) and Ansorg et al. (2002b).

A4.2 Generalization of the Bäcklund type solutions by a limiting process

For a given function ξ, one can use Equation (A4.5) to write

$$\exp\left[x\,\xi(x^2)\right] \approx \prod_{\nu=1}^{q} \frac{i X_\nu - x}{i X_\nu + x} = \frac{P_q(-x)}{P_q(x)} \qquad (A4.7)$$

with

$$P_q(x) = \sum_{\nu=0}^{q-1} b_\nu x^\nu + x^q.$$

The coefficients b_ν can be determined by equating left and right hand sides of (A4.7) at the q zeros $x_\mu^2 \in [0, 1]$ of the Chebyshev polynomial $T_q(2x^2 - 1)$ and solving the corresponding linear system:

$$\exp\left[x_\mu \xi(x_\mu^2)\right] \sum_{\nu=0}^{q} b_\nu x_\mu^\nu = \sum_{\nu=0}^{q} b_\nu (-x_\mu)^\nu.$$

The zeros of P_q determine the X_ν. In the limit $q \to \infty$, we thus obtain an exact representation of ξ in $[0, 1]$.

References

Abramowicz, M. A., Calvani, M. and Nobili, L. (1983). Runaway instability in accretion disks orbiting black holes. *Nature*, **302**, 597.

Abramowicz, M. A., Curir, A., Schwarzenberg-Czerny, A. and Wilson, R. E. (1984). Self-gravity and the global structure of accretion discs. *Mon. Not. R. Astron. Soc.*, **208**, 279.

Ansorg, M. (1998). Timelike geodesic motions within the general relativistic gravitational field of the rigidly rotating disk of dust. *J. Math. Phys.*, **39**, 5984.

Ansorg, M. (2001). Differentially rotating disks of dust: Arbitrary rotation law. *Gen. Rel. Grav.*, **33**, 309.

Ansorg, M., Fischer, T., Kleinwächter, A., Meinel, R., Petroff, D. and Schöbel, K. (2004). Equilibrium configurations of homogeneous fluids in general relativity. *Mon. Not. R. Astron. Soc.*, **355**, 682.

Ansorg, M., Kleinwächter, A. and Meinel, R. (2002a). Highly accurate calculation of rotating neutron stars. *Astron. Astrophys.*, **381**, L49.

Ansorg, M., Kleinwächter, A. and Meinel, R. (2003a). Highly accurate calculation of rotating neutron stars: Detailed description of the numerical methods. *Astron. Astrophys.*, **405**, 711.

Ansorg, M., Kleinwächter, A. and Meinel, R. (2003b). Relativistic Dyson rings and their black hole limit. *Astrophys. J. Lett.*, **582**, L87.

Ansorg, M., Kleinwächter, A. and Meinel, R. (2003c). Uniformly rotating axisymmetric fluid configurations bifurcating from highly flattened Maclaurin spheroids. *Mon. Not. R. Astron. Soc.*, **339**, 515.

Ansorg, M., Kleinwächter, A., Meinel, R. and Neugebauer, G. (2002b). Dirichlet boundary value problems of the Ernst equation. *Phys. Rev. D*, **65**, 044006.

Ansorg, M. and Petroff, D. (2005). Black holes surrounded by uniformly rotating rings. *Phys. Rev. D*, **72**, 024019.

Ansorg, M. and Petroff, D. (2006). Negative Komar mass of single objects in regular, asymptotically flat spacetimes. *Class. Quantum Grav.*, **23**, L81.

Ansorg, M. and Pfister, H. (2008). A universal constraint between charge and rotation rate for degenerate black holes surrounded by matter. *Class. Quantum Grav.*, **25**, 035009.

Arnold, D. N. (2001). *A Concise Introduction to Numerical Analysis* (Minneapolis, Institute for Mathematics and its Applications, University of Minnesota). URL www.ima.umn.edu/ arnold/597.00-01/nabook.pdf.

Ashtekar, A. and Krishnan, B. (2004). Isolated and dynamical horizons and their applications. *Living Reviews in Relativity*, **7**. Cited on May 12, 2007, URL `www.livingreviews.org/lrr-2004-10`.

Bardeen, J. M. (1970). A variational principle for rotating stars in General Relativity. *Astrophys. J.*, **162**, 71.

Bardeen, J. M. (1971). A reexamination of the post-Newtonian Maclaurin spheroids. *Astrophys. J.*, **167**, 425.

Bardeen, J. M. (1973a). Rapidly rotating stars, disks, and black holes. In C. DeWitt and B. S. DeWitt (eds.), *Black Holes, Les astres occlus* (New York, Gordon and Breach Science Publishers), pp. 241–289.

Bardeen, J. M. (1973b). Timelike and null geodesics in the Kerr metric. In C. DeWitt and B. S. DeWitt (eds.), *Black Holes, Les astres occlus* (New York, Gordon and Breach Science Publishers), pp. 215–240.

Bardeen, J. M. and Horowitz, G. T. (1999). Extreme Kerr throat geometry: A vacuum analog of $AdS_2 \times S^2$. *Phys. Rev. D*, **60**, 104030.

Bardeen, J. M., Press, W. H. and Teukolsky, S. A. (1972). Rotating black holes: Locally nonrotating frames, energy extraction, and scalar synchrotron radiation. *Astrophys. J.*, **178**, 347.

Bardeen, J. M. and Wagoner, R. V. (1971). Relativistic disks. I. Uniform rotation. *Astrophys. J.*, **167**, 359.

Beig, R. and Chruściel, P. (2006). Stationary black holes. In J.-P. Françoise, G. L. Naber and Tsou S. T. (eds.), *Encyclopedia of Mathematical Physics*, vol. 5 (Oxford, Academic Press/Elsevier), pp. 38–44. E-print: `gr-qc/0502041`.

Belinski, V. A. and Zakharov, V. E. (1978). Integration of the Einstein equations by the inverse scattering problem technique and the calculation of the exact soliton solutions. *Zh. Eksp. Teor. Fiz. Pis'ma*, **75**, 195.

Belokolos, E. D., Bobenko, A. I., Enol'skii, V. Z., Its, A. R. and Matveev, V. B. (1994). *Algebro-Geometric Approach to Nonlinear Integrable Equations*. Nonlinear Dynamics (New York, Springer).

Birkhoff, G. and Lynch, R. E. (1984). *Numerical Solution of Elliptic Problems*. Studies in Applied Mathematics (Philadelphia, SIAM).

Bodo, G. and Curir, A. (1992). Models of self-gravitating accretion disks. *Astron. Astrophys.*, **253**, 318.

Bonazzola, S., Gourgoulhon, E. and Marck, J.-A. (1998). Numerical approach for high precision 3-D relativistic star models. *Phys. Rev. D*, **58**, 104020.

Bonazzola, S., Gourgoulhon, E., Salgado, M. and Marck, J.-A. (1993). Axisymmetric rotating relativistic bodies: A new numerical approach for 'exact' solutions. *Astron. Astrophys.*, **278**, 421.

Bonazzola, S. and Schneider, J. (1974). An exact study of rigidly and rapidly rotating stars in general relativity with application to the Crab pulsar. *Astrophys. J.*, **191**, 273.

Boyer, R. H. and Lindquist, R. W. (1967). Maximal analytic extension of the Kerr metric. *J. Math. Phys.*, **8**, 265.

Buchdahl, H. A. (1959). General relativistic fluid spheres. *Phys. Rev.*, **116**, 1027.

Butterworth, E. M. and Ipser, J. R. (1976). On the structure and stability of rapidly rotating fluid bodies in general relativity. I. The numerical method for computing structure and its application to uniformly rotating homogeneous bodies. *Astrophys. J.*, **204**, 200.

Carter, B. (1973). Black hole equilibrium states. In C. DeWitt and B. S. DeWitt (eds.), *Black Holes, Les astres occlus* (New York, Gordon and Breach Science Publishers), pp. 57–214.

Chandrasekhar, S. (1939). *An Introduction to the Study of Steller Structure* (Chicago, University of Chicago Press).
Chandrasekhar, S. (1967). The post-Newtonian effects of General Relativity on the equilibrium of uniformly rotating bodies II. The deformed figures of the Maclaurin spheroids. *Astrophys. J.*, **147**, 334.
Chandrasekhar, S. (1968). The virial equations of the fourth order. *Astrophys. J.*, **152**, 293.
Chandrasekhar, S. (1969). *Ellipsoidal Figures of Equilibrium* (New Haven, Yale University Press). A revised edition was published by Dover, New York, in 1987.
Chandrasekhar, S. (1970). Solutions of two problems in the theory of gravitational radiation. *Phys. Rev. Lett.*, **24**, 611.
Chandrasekhar, S. and Friedman, J. L. (1972a). On the stability of axisymmetric systems to axisymmetric perturbations in general relativity. I. The equations governing nonstationary, and perturbed systems. *Astrophys. J.*, **175**, 379.
Chandrasekhar, S. and Friedman, J. L. (1972b). On the stability of axisymmetric systems to axisymmetric perturbations in general relativity. II. A criterion for the onset of instability in uniformly rotating configurations and the frequency of the fundamental mode in case of slow rotation. *Astrophys. J.*, **176**, 745.
Chandrasekhar, S. and Friedman, J. L. (1973). On the stability of axisymmetric systems to axisymmetric perturbations in general relativity. IV. Allowance for gravitational radiation in an odd-parity mode. *Astrophys. J.*, **181**, 481.
Christodoulou, D. (1970). Reversible and irreversible transformations in black-hole physics. *Phys. Rev. Lett.*, **25**, 1596.
Chruściel, P. T., Greuel, G.-M., Meinel, R. and Szybka, S. J. (2006). The Ernst equation and ergosurfaces. *Class. Quantum Grav.*, **23**, 4399.
Contopoulos, G. (1994). Order and chaos. In G. Contopoulos, N. K. Spyrou and L. Vlahos (eds.), *Galactic Dynamics and N-Body Simulations, Lecture Notes in Physics*, vol. 433 (Berlin, Springer Verlag), pp. 33–100.
Cook, G. B., Shapiro, S. L. and Teukolsky, S. A. (1992). Spin-up of a rapidly rotating star by angular momentum loss: Effects of general relativity. *Astrophys. J.*, **398**, 203.
Detweiler, S. L. and Lindblom, L. (1977). On the evolution of the homogeneous ellipsoidal figures. *Astrophys. J.*, **213**, 193.
Dyson, F. W. (1892). The potential of an anchor ring. *Phil. Trans. R. Soc. London, Ser. A*, **184**, 43.
Dyson, F. W. (1893). The potential of an anchor ring – part II. *Phil. Trans. R. Soc. London, Ser. A*, **184**, 1041.
Eisenstat, S. C. (1974). On the rate of convergence of the Bergman–Vekua method for the numerical solution of elliptic boundary problems. *SIAM J. Numer. Anal.*, **11**, 654.
Eriguchi, Y. and Hachisu, I. (1982). New equilibrium sequences bifurcating from Maclaurin sequence. *Prog. Theor. Phys.*, **67**, 844.
Eriguchi, Y. and Sugimoto, D. (1981). Another equilibrium sequence of self-gravitating and rotating incompressible fluid. *Prog. Theor. Phys.*, **65**, 1870.
Ernst, F. (1968). New formulation of the axially symmetric gravitational field problem. *Phys. Rev*, **167**, 1175.
Ernst, F. (1977). A new family of solutions of the Einstein field equations. *J. Math. Phys.*, **18**, 233.
Fischer, T., Horatschek, S. and Ansorg, M. (2005). Uniformly rotating rings in general relativity. *Mon. Not. R. Astron. Soc.*, **364**, 943.
Fodor, G., Hoenselaers, C. and Perjés, Z. (1989). Multipole moments of axisymmetric systems in relativity. *J. Math. Phys.*, **30**, 2252.

Font, J. A. (2003). Numerical hydrodynamics in general relativity. *Living Reviews in Relativity*, **6**. Cited on May 12, 2007, URL www.livingreviews.org/lrr-2003-04.
Friedman, J. L. and Ipser, J. R. (1992). Rapidly rotating relativistic stars. *Phil. Trans. R. Soc. Lond. A*, **340**, 391.
Friedman, J. L., Ipser, J. R. and Parker, L. (1986). Rapidly rotating neutron star models. *Astrophys. J.*, **304**, 115. Erratum (1990): *Astrophys. J.*, **351**, 705.
Friedman, J. L., Ipser, J. R. and Parker, L. (1989). Implications of a half-millisecond pulsar. *Phys. Rev. Lett.*, **62**, 3015.
Friedman, J. L., Ipser, J. R. and Sorkin, R. D. (1988). Turning-point method for axisymmetric stability of rotating relativistic stars. *Astrophys. J.*, **325**, 722.
Friedman, J. L. and Schutz, B. F. (1978). Secular instability of rotating Newtonian stars. *Astrophys. J.*, **222**, 281.
Geroch, R. (1970). Multipole moments. II. Curved space. *J. Math. Phys.*, **11**, 2580.
Göpel, A. (1847). Entwurf einer Theorie der Abel'schen Transcendenten erster Ordnung. *Crelle's J. für Math.*, **35**, 227.
Gourgoulhon, E., Haensel, P., Livine, R., Paluch, E., Bonazzola, S. and Marck, J.-A. (1999). Fast rotation of strange stars. *Astron. Astrophys.*, **349**, 851.
Hachisu, I. and Eriguchi, Y. (1984). Bifurcation points on the Maclaurin sequence. *Publ. Astron. Soc. Japan*, **36**, 497.
Hansen, R. O. (1974). Multipole moments of stationary space-times. *J. Math. Phys.*, **15**, 46.
Harrison, B. K. (1978). Bäcklund transformation for the Ernst equation of general relativity. *Phys. Rev. Lett.*, **41**, 1197.
Hartle, J. B. and Sharp, D. H. (1967). Variational principle for the equilibrium of a relativistic, rotating star. *Astrophys. J.*, **147**, 317.
Hawking, S. and Ellis, G. (1973). *The Large Scale Structure of Space-Time* (Cambridge, Cambridge University Press).
Heusler, M. (1996). *Black Hole Uniqueness Theorems* (Cambridge, Cambridge University Press).
Hoenselaers, C., Kinnersley, W. and Xanthopoulos, B. C. (1979). Generation of asymptotically flat, stationary space-times with any number of parameters. *Phys. Rev. Lett.*, **42**, 481.
Horatschek, S. and Petroff, D. (2008). Uniformly rotating homogeneous rings in Newtonian gravity. E-print: arXiv: 0802.0078.
Kerr, R. P. (1963). Gravitational field of a spinning mass as an example of algebraically special metrics. *Phys. Rev. Lett.*, **11**, 237.
Kippenhahn, R. and Weigert, A. (1990). *Stellar Structure and Evolution* (Berlin, Springer-Verlag).
Kleinwächter, A. (2000). Properties of the Neugebauer–Meinel solution. *Ann. Phys. (Leipzig)*, **9**, 99. Special Issue.
Kleinwächter, A. (2001). Discussion of the theta formula for the Ernst potential of the rigidly rotating disk of dust. In A. Macias, J. L. Cervantes-Cota and C. Lämmerzahl (eds.), *Exact Solutions and Scalar Fields in Gravity* (Kluwer Press, New York), pp. 39–51.
Kleinwächter, A., Meinel, R. and Neugebauer, G. (1995). The multipole moments of the rigidly rotating disk of dust in general relativity. *Phys. Lett. A*, **200**, 82.
Komar, A. (1959). Covariant conservation laws in general relativity. *Phys. Rev.*, **113**, 934.

Komatsu, H., Eriguchi, Y. and Hachisu, I. (1989a). Rapidly rotating general relativistic stars – I. Numerical method and its application to uniformly rotating polytropes. *Mon. Not. R. Astron. Soc.*, **237**, 355.

Komatsu, H., Eriguchi, Y. and Hachisu, I. (1989b). Rapidly rotating general relativistic stars – II. Differentially rotating polytropes. *Mon. Not. R. Astron. Soc.*, **239**, 153.

Kordas, P. (1995). Reflection-symmetric, asymptotically flat solutions of the vacuum axistationary Einstein equations. *Class. Quantum Grav.*, **12**, 2037.

Kowalewsky, S. (1885). Zusätze und Bemerkungen zu Laplace's Untersuchung über die Gestalt der Saturnringe. *Astronomische Nachrichten*, **111**, 37.

Kramer, D. and Neugebauer, G. (1968). Zu axialsymmetrischen stationären Lösungen der Einsteinschen Feldgleichungen für das Vakuum. *Commun. Math. Phys.*, **10**, 132.

Krazer, A. (1903). *Lehrbuch der Thetafunktionen* (Leipzig, Teubner).

Kundt, W. and Trümper, M. (1966). Orthogonal decomposition of axi-symmetric stationary spacetimes. *Z. Phys.*, **192**, 419.

Labranche, H., Petroff, D. and Ansorg, M. (2007). The quasi-stationary transition of strange matter rings to a black hole. *Gen. Rel. Grav.*, **39**, 129.

Landau, L. D. (1932). On the theory of stars. *Phys. Z. Sowjetunion*, **1**, 285.

Landau, L. D. and Lifshitz, E. M. (1980). *Statistical Physics: Part 1* (Oxford, Pergamon Press).

Lanza, A. (1992). Self-gravitating thin disks around rapidly rotating black holes. *Astrophys. J.*, **389**, 141.

Lattimer, J. M., Prakash, M., Masak, D. and Yahil, A. (1990). Rapidly rotating pulsars and the equation of state. *Astrophys. J.*, **355**, 241.

Lewis, T. (1932). Some special solutions of the equations of axially symmetric gravitational fields. *Proc. Roy. Soc. Lond. A*, **136**, 176.

Lichtenstein, L. (1933). *Gleichgewichtsfiguren rotierender Flüssigkeiten* (Berlin, Springer).

Lindblom, L. (1992). On the symmetries of equilibrium stellar models. *Phil. Trans. R. Soc. Lond. A*, **340**, 353.

Lindblom, L. and Hiscock, W. A. (1983). On the stability of rotating stellar models in general relativity theory. *Astrophys. J.*, **267**, 384.

Maclaurin, C. (1742). *A Treatise on Fluxions* (Edinburgh, Ruddimans).

Maison, D. (1978). Are the stationary, axially symmetric Einstein equations completely integrable? *Phys. Rev. Lett.*, **41**, 521.

Masood-ul-Alam, A. K. M. (2007). Proof that static stellar models are spherical. *Gen. Rel. Grav.*, **39**, 55.

Meinel, R. (2002). Black holes: A physical route to the Kerr metric. *Ann. Phys. (Leipzig)*, **11**, 509.

Meinel, R. (2004). Quasistationary collapse to the extreme Kerr black hole. *Ann. Phys. (Leipzig)*, **13**, 600.

Meinel, R. (2006). On the black hole limit of rotating fluid bodies in equilibrium. *Class. Quantum Grav.*, **23**, 1359.

Meinel, R. and Kleinwächter, A. (1995). Dragging effects near a rigidly rotating disk of dust. In J. B. Barbour and H. Pfister (eds.), *Mach's Principle: From Newton's Bucket to Quantum Gravity* (Boston, Birkhäuser), pp. 339–346.

Meinel, R. and Neugebauer, G. (1995). Asymptotically flat solutions to the Ernst equation with reflection symmetry. *Class. Quantum Grav.*, **12**, 2045.

Montero, P. J., Rezzolla, L. and Yoshida, S. (2004). Oscillations of vertically integrated relativistic tori – II. Axisymmetric modes in a Kerr spacetime. *Mon. Not. R. Astron. Soc.*, **354**, 1040.

Neugebauer, G. (1979). Bäcklund transformations of axially symmetric stationary gravitational fields. *J. Phys. A*, **12**, L67.
Neugebauer, G. (1980a). A general integral of the axially symmetric stationary Einstein equations. *J. Phys. A*, **13**, L19.
Neugebauer, G. (1980b). Recursive calculation of axially symmetric stationary Einstein fields. *J. Phys. A*, **13**, 1737.
Neugebauer, G. (1988). Thermodynamics of rotating bodies. In Z. Perjés (ed.), *Relativity Today* (Singapore, World Scientific), pp. 134–146.
Neugebauer, G. (1996). Gravitostatics and rotating bodies. In G. S. Hall and J. R. Pulham (eds.), *General Relativity* (London, SUSSP Publications and Institute of Physics Publishing), pp. 61–81.
Neugebauer, G. (2000). Rotating bodies as boundary value problems. *Ann. Phys. (Leipzig)*, **9**, 342.
Neugebauer, G. and Herold, H. (1992). Gravitational fields of rapidly rotating neutron stars: Theoretical foundation. In J. Ehlers and G. Schäfer (eds.), *Relativistic Gravity Research: With Emphasis on Experiments and Observations, Lecture Notes in Physics*, vol. 410 (Berlin, Springer), pp. 305–318.
Neugebauer, G., Kleinwächter, A. and Meinel, R. (1996). Relativistically rotating dust. *Helv. Phys. Acta*, **69**, 472.
Neugebauer, G. and Kramer, D. (1983). Einstein–Maxwell solitons. *J. Phys. A*, **16**, 1927.
Neugebauer, G. and Meinel, R. (1993). The Einsteinian gravitational field of a rigidly rotating disk of dust. *Astrophys. J.*, **414**, L97.
Neugebauer, G. and Meinel, R. (1994). General relativistic gravitational field of a rigidly rotating disk of dust: Axis potential, disk metric and surface mass density. *Phys. Rev. Lett.*, **73**, 2166.
Neugebauer, G. and Meinel, R. (1995). General relativistic gravitational field of a rigidly rotating disk of dust: Solution in terms of ultraelliptic functions. *Phys. Rev. Lett.*, **75**, 3046.
Neugebauer, G. and Meinel, R. (2003). Progress in relativistic gravitational theory using the inverse scattering method. *J. Math. Phys.*, **44**, 3407.
Nishida, S. and Eriguchi, Y. (1994). A general relativistic toroid around a black hole. *Astrophys. J.*, **427**, 429.
Novikov, S., Manakov, S. V., Pitaevskii, L. P. and Zakharov, V. E. (1984). *Theory of Solitons: The Inverse Scattering Method* (New York, Consultants Bureau).
Nozawa, T., Stergioulas, N., Gourgoulhon, E. and Eriguchi, Y. (1998). Construction of highly accurate models of rotating neutron stars – comparison of three different numerical schemes. *Astron. Astrophys. Supp.*, **132**, 431.
Oppenheimer, J. R. and Volkoff, G. M. (1939). On massive neutron cores. *Phys. Rev.*, **55**, 374.
Ostriker, J. (1964). The equilibrium of self-gravitating rings. *Astrophys. J.*, **140**, 1067.
Papapetrou, A. (1966). Champs gravitationnels stationnaires à symétrie axiale. *Ann. Inst. H. Poincaré A*, **4**, 83.
Peshier, A., Kämpfer, B. and Soff, G. (2000). The equation of state of deconfined matter at finite chemical potential in a quasiparticle description. *Phys. Rev. C*, **61**, 045203.
Petroff, D. (2003). Post-Newtonian Maclaurin spheroids to arbitrary order. *Phys. Rev. D*, **68**, 104029.
Petroff, D. and Horatschek, S. (2008). Uniformly rotating polytropic rings in Newtonian gravity. E-print: `arXiv: 0802.0081`.

Petroff, D. and Meinel, R. (2001). Post-Newtonian approximation of the rigidly rotating disk of dust to arbitrary order. *Phys. Rev. D*, **63**, 064012.

Poincaré, H. (1885). Sur l'équilibre d'une masse fluide animée d'un mouvement de rotation. *Acta mathematica*, **7**, 259.

Rezzolla, L., Yoshida, S. and Zanotti, O. (2003). Oscillations of vertically integrated relativistic tori – I. Axisymmetric modes in a Schwarzschild spacetime. *Mon. Not. R. Astron. Soc.*, **344**, 978.

Robinson, D. C. (1975). Uniqueness of the Kerr black hole. *Phys. Rev. Lett.*, **34**, 905.

Roche, É. (1873). Essai sur la constitution et l'origine du système solaire. *Mém. de la section des sciences, Acad. des sciences et lettres de Montpellier*, **1**, 235.

Rosenhain, G. (1850). Auszug mehrerer Schreiben des Herrn Dr. Rosenhain an Herrn Prof. C.G.J. Jacobi über die hyperelliptischen Transcendenten. *Crelle's J. für Math.*, **40**, 319.

Schöbel, K. and Ansorg, M. (2003). Maximal mass of uniformly rotating homogeneous stars in Einsteinian gravity. *Astron. Astrophys.*, **405**, 405.

Schoen, R. and Yau, S.-T. (1979). Positivity of the total mass of a general space-time. *Phys. Rev. Lett.*, **43**, 1457.

Schutz, B. F. (1972). Linear pulsations and stability of differentially rotating stellar models. II. General-relativistic analysis. *Astrophys. J. Supp.*, **24**, 343.

Schwarzschild, K. (1916). Über das Gravitationsfeld einer Kugel aus inkompressibler Flüssigkeit nach der Einsteinschen Theorie. *Sitzungsberichte der Königlich Preussischen Akademie der Wissenschaften*, **1**, 424.

Sen, N. R. (1934). On the equilibrium of an incompressible sphere. *Mon. Not. R. Astron. Soc.*, **94**, 550.

Shapiro, S. L. and Shibata, M. (2002). Collapse of a rotating supermassive star to a supermassive black hole: Analytic determination of the black hole mass and spin. *Astrophys. J.*, **577**, 904.

Shapiro, S. L. and Teukolsky, S. A. (1983). *Black Holes, White Dwarfs, and Neutron Stars. The Physics of Compact Objects* (New York, John Wiley).

Shibata, M., Taniguchi, K. and Uryu, K. (2003). Merger of binary neutron stars of unequal mass in full general relativity. *Phys. Rev. D*, **68**, 084020.

Smarr, L. (1973). Mass formula for Kerr black holes. *Phys. Rev. Lett.*, **30**, 71.

Stahl, H. (1896). *Theorie der Abel'schen Funktionen* (Leipzig, Teubner).

Stephani, H. (2004). *Relativity* (Cambridge, Cambridge University Press).

Stephani, H., Kramer, D., MacCallum, M., Hoenselaers, C. and Herlt, E. (2003). *Exact Solutions of Einstein's Field Equations*, 2nd edn. (Cambridge, Cambridge University Press).

Stergioulas, N. (2003). Rotating stars in relativity. *Living Reviews in Relativity*, **6**. Cited on 11 April 2007, URL www.livingreviews.org/lrr-2003-3.

Stergioulas, N. and Friedman, J. L. (1995). Comparing models of rapidly rotating relativistic stars constructed by two numerical methods. *Astrophys. J.*, **444**, 306.

Stoner, E. C. (1932). The minimum pressure of a degenerate electron gas. *Mon. Not. R. Astron. Soc.*, **92**, 651.

Thorne, K. S. (1967). The general relativistic theory of stellar structure and dynamics. In C. DeWitt, E. Schatzman and P. Véron (eds.), *High Energy Astrophysics, Volume 3* (New York, Gordon and Breach), pp. 259–441.

Thorne, K. S. (1980). Multipole expansions of gravitational radiation. *Rev. Mod. Phys.*, **52**, 299.

Tolman, R. C. (1934). *Relativity, Thermodynamics and Cosmology* (Oxford, Oxford University Press).

Tolman, R. C. (1939). Static solutions of Einstein's field equations for spheres of fluid. *Phys. Rev.*, **55**, 364.

Tooper, R. F. (1964). General relativistic polytropic fluid spheres. *Astrophys. J.*, **140**, 434.

Tooper, R. F. (1965). Adiabatic fluid spheres in general relativity. *Astrophys. J.*, **142**, 1541.

Trümper, M. (1967). Einsteinsche Feldgleichungen für das axialsymmetrische, stationäre Gravitationsfeld im Innern einer starr rotierenden Flüssigkeit. *Z. Naturforsch.*, **22a**, 1347.

Wald, R. M. (1984). *General Relativity* (Chicago, University of Chicago Press).

Wigley, N. M. (1970). Mixed boundary value problems in plane domains with corners. *Math. Z.*, **115**, 33.

Will, C. M. (1974). Perturbation of a slowly rotating black hole by a stationary axisymmetric ring of matter. I. Equilibrium configurations. *Astrophys. J.*, **191**, 521.

Will, C. M. (1975). Perturbation of a slowly rotating black hole by a stationary axisymmetric ring of matter. II. Penrose processes, circular orbits and differential mass formulæ. *Astrophys. J.*, **196**, 41.

Wilson, J. R. (1972). Models of differentially rotating stars. *Astrophys. J.*, **176**, 195.

Wolf, T. (1998). Structural equations for Killing tensors of arbitrary rank. *Computer Phys. Commun.*, **115**, 316.

Wong, C. Y. (1974). Toroidal figures of equilibrium. *Astrophys. J.*, **190**, 675.

Zanotti, O., Font, J. A., Rezzolla, L. and Montero, P. J. (2005). Dynamics of oscillating relativistic tori around Kerr black holes. *Mon. Not. R. Astron. Soc.*, **356**, 1371.

Zel'dovich, Y. B. and Novikov, I. D. (1971). *Relativistic Astrophysics*, vol. 1 (Chicago, University of Chicago Press).

Index

4-acceleration, 24
4-velocity, 5, 24

Abelian differential, 55
accretion discs, 176
angular momentum, 2, 13
    disc of dust, 79
    Kerr, 27, 112
    Maclaurin disc, 38
    Maclaurin spheroid, 36
angular velocity, 5
    Maclaurin disc, 38
    of locally non-rotating observer, 91
    of the horizon, 26, 28, 109, 112
asymptotic behaviour, 13, 36
asymptotic flatness, 2
axial symmetry, 3
axis of symmetry, 4, 42

Bäcklund transformation, 112, 204
Bernoulli–l'Hospital rule, 112
bifurcation points, 139
binding energy, 157, 160
    disc of dust, 87, 95
black hole, 26, 31, 108
    degenerate, 176
    extreme Kerr, 29, 31, 146, 152
    Kerr, 27
    Schwarzschild, 29
    surrounded by a fluid ring, 166
black hole limit, 31, 147
    disc of dust, 104
    rings, 165
black hole uniqueness, 33, 112
boundary conditions
    horizon, 27, 167
boundary of the fluid body, 6
boundary value problem, 40
    black hole, 109
    disc, 22
Boyer–Lindquist coordinates, 27
branch cut, 43
branch points, 43
    confluent, 43
Buchdahl limit, 18, 39, 141

centrifugal force, 25, 38
centrifugal potential, 17
CFS instability, 180
Chebyshev coefficients, 122
Chebyshev expansion, 116
Chebyshev polynomials, 116
chemical potential, 13
Christoffel symbols, 24
circular orbits, 28, 91, 97
circumferential radius, 149
corotating potentials, 5, 8, 17
corotating system, 5, 10, 15, 45, 80, 109
collapse, 19, 166
collocation point, 121, 135
compactification, 116, 129, 132
conformal transformation, 9
coordinate mappings, 127
covariant derivative, 6
cylindrical coordinates, 5

Dirac delta distribution, 20
direct orbits, 28, 91
disc limit, 19, 153
disc of dust, 25, 38, 40, 153
dust limit, 12, 25
Dyson rings, 144

Einstein's field equations, 7
    vacuum, 10
elliptic coordinates
    oblate, 34
elliptic functions, 70, 71, 86, 190
elliptic integrals, 86, 190
embedding diagram, 30
energy-density, 6, 10
    internal, 10
energy-momentum tensor
    disc, 24
    dust, 12
    perfect fluid, 6
enthalpy, 6, 13, 115

Index

equation of state, 6, 10
  barotropic, 153
  completely degenerate, ideal gas
    of neutrons, 11, 161
  homogeneous fluid, 10, 137
  polytropic, 11, 154
  strange quark matter, 12, 162
equator, 25
equatorial plane, 28
ergosphere, 14, 141, 144
  disc of dust, 88
  Kerr black hole, 28
ergosurface, 14
Ernst equation, 2, 10, 42, 109
  corotating, 23
Ernst potential, 10, 82, 204
  Kerr solution, 112
Euler equation, 17, 36
extreme Kerr solution, 112

far field, 1
Fermi gas
  completely degenerate, ideal, 11, 161
Fermi–Dirac statistics
  special-relativistic, 11
field equations, 7, 114
frame dragging, 171
free boundary value problem, 128

Gauss' theorem, 20
general relativity, 17
geodesic motion, 12, 25, 28, 91, 96
Gibbs phenomenon, 119, 120
global problem, 10
gravitational energy, 37
  Maclaurin disc, 38
gravitational radiation, 3
gravitomagnetic effects, 2, 166
gravitomagnetic potential, 4

Hamiltonian system, 96
Heuman's lambda function, 58, 187
holomorphic function, 47
homogeneous fluids, 10
horizon, 26, 31, 109, 167
  area, 175
  degenerate, 33, 112, 168
  dynamical, 176
  isolated, 176
hyperelliptic functions, 54
hypersurface
  null, 26, 32
  spacelike, 13
  timelike, 22

infinity, 2, 5, 7
  spatial, 5, 42
integrability condition, 9, 42
inverse method, 41, 108
  direct problem, 42

isolated body, 1
isotropic coordinates, 39

Jacobi's inversion problem, 54
Jacobi's zeta function, 58, 86, 187
Jacobian matrix, 123
jump matrix, 49

Kerr metric, 27, 108
Kerr solution, 27, 112
  extreme, 107, 113
Killing vector, 4, 14, 26
kinetic energy
  Maclaurin disc, 38
  of rotation, 36
Komar mass, 169
  individual, 170
  total, 170
Kronecker symbol, 117

Laplace equation, 16, 36, 182
  two-dimensional, 9, 21
Legendre functions, 124, 139
Legendre polynomials, 139
Lense–Thirring effect, 2
Lewis–Papapetrou metric, 4
line element, 4, 114
  far field, 1
line integration, 23
linear problem, 41
local inertial system, 2
locally non-rotating observers, 8, 91
Lyapunov functional, 179

Maclaurin disc, 37
Maclaurin sequence, 140
Maclaurin spheroid, 34
marginally bound orbit, 28
marginally stable orbit, 28
mass
  baryonic, 13, 40, 79
  Christodoulou, 175, 176
  disc of dust, 79
  gravitational, 2, 13, 39, 79
  Kerr, 27, 112
  Komar, 169
  Maclaurin spheroid, 36
mass-density, 6, 10
  baryonic, 6, 10
mass-shed parameter, 141
mass-shedding limit, 25, 141, 146, 150, 181
matching conditions, 2
metric, 114
  asymptotic behaviour, 14
  axisymmetric perfect fluid body in
    stationary rotation, 4
  disc of dust, 57
  far field, 1
  Lewis–Papapetrou, 4
  Minkowski, 2, 5, 16
MIT bag constant, 12

MIT bag model, 12, 163
moment of inertia, 37
multipole moments, 82, 107, 193

near-horizon geometry, 108
neutron gas, 161
neutron stars, 3, 11
Newton–Raphson scheme, 123
Newtonian limit, 16
   disc of dust, 52, 93
   Schwarzschild spheres, 40
Newtonian potential, 16, 34, 181
   generalized, 4, 24
non-rotating limit, 17
normal vector, 13, 22, 26, 33

orthogonal transitivity, 4, 33

partial derivatives, 8
perfect fluid, 3, 6
photon orbit, 28
Poisson equation, 16, 36, 182
Poisson integral, 34
polytrope, 11, 157
polytropic constant, 11, 154
polytropic exponent, 11
polytropic index, 11, 154
positive mass theorem, 170
post-Newtonian expansion
   disc limit, 19
   disc of dust, 95
   Maclaurin spheroids, 139
pressure, 6
pseudo-spectral method, 115

radius ratio, 135
redshift, 7
reflectional symmetry, 20, 25, 83, 204
retrograde orbits, 28, 91
Riemann matrix, 56
Riemann surface, 43, 54
Riemann–Hilbert problem, 49, 51
rigid rotation, 3, 5
rings, 145
   black hole limit, 165
Roche model, 184
Rosenhain's theta functions, 67, 76, 189
rotation
   rigid, 3, 5
   with respect to infinity, 2, 5, 175
   with respect to the 'fixed stars', 2
   with respect to the local inertial system, 2

Schwarzschild coordinates, 18
Schwarzschild metric, 39
Schwarzschild solution, 112, 141
   interior, 39
seed solution, 203
soliton theory, 2, 41
spectral approximation, 119

spectral coefficients, 117
spectral expansion, 116
spectral parameter, 41
spectral resolution, 119
speed of light, 92
speed of sound, 11
spherical symmetry, 18
spheroidal configurations, 129
stability, 3, 177
   dynamical, 177, 179
   secular, 137, 177
static model, 17
stationarity, 3
   local, 14
stationary limit, 15
strange matter, 12, 162
strange quark matter, 6, 12
subdomains, 128
superluminal motion, 15
surface
   of the fluid, 7
   shape, 10, 34
surface condition, 6, 31, 35
   Newtonian, 17
surface energy-density, 22
surface gravity, 112, 167
surface layer, 20
surface mass-density, 79
   Maclaurin disc, 38
symmetry axis, 4

temperature, 3, 6
thermodynamic equilibrium, 3
theta functions, 54, 57, 187
   elliptic, 58, 70
   hyperelliptic, 187
   moduli, 59
   ultra-elliptic, 58, 187
throat geometry, 30, 33, 108
Tolman condition, 3, 13
Tolman–Oppenheimer–Volkoff equation, 18
topology, 146, 169
toroidal configurations, 130
two-body limit, 153
two-body systems
   stationary, 166

variational principle, 13
velocity of rotation, 8, 16, 91
viscosity, 3, 177
volume element, 13

Weierstrass function, 105
Weyl coordinates
   canonical, 9, 21, 109
Weyl–Lewis–Papapetrou coordinates, 112
white dwarfs, 11

zero angular momentum observers, 8, 167